# WRITING FOR TECHNICAL AND BUSINESS MAGAZINES

WILEY SERIES ON HUMAN COMMUNICATION

W. A. Mambert
*PRESENTING TECHNICAL IDEAS: A Guide to Audience Communication*

William J. Bowman
*GRAPHIC COMMUNICATION*

Herman M. Weisman
*TECHNICAL CORRESPONDENCE: A Handbook and Reference Source for the Technical Professional*

John H. Mitchell
*WRITING FOR TECHNICAL AND PROFESSIONAL JOURNALS*

James J. Welsh
*THE SPEECH WRITING GUIDE: Professional Techniques for Regular and Occasional Speakers*

Michael P. Jaquish
*PERSONAL RESUME PREPARATION*

George T. Vardaman and Carroll C. Halterman
*COMMUNICATION FOR MANAGERIAL CONTROL: Systems for Organizational Diagnosis and Design*

John D. Haughney
*EFFECTIVE CATALOGS*

# WRITING FOR TECHNICAL AND BUSINESS MAGAZINES

ROBERT H. DODDS, P.E.

JOHN WILEY & SONS, INC.

New York · London · Sydney · Toronto

10 9 8 7 6 5 4 3 2 1
Library of Congress Catalogue Card Number: 70-93486

SBN 471 21725 5

Printed in the United States of America

To

John Simpson Dodds

# PREFACE

This book is half of a book idea that became two books. Consideration of the reader, factor No. 1 in effective writing, fostered the split.

Here's the story. Professor John H. Mitchell, of the University of Massachusetts, agreed to do a book for the Wiley-Interscience audience on writing for professional publications. His analysis of the assignment led him to the conclusion that papers for journals and articles for magazines are two separate subjects for two different audiences—even though many readers would be authors of both journal papers and magazine articles. Professor Mitchell chose to proceed with *Writing for Technical and Professional Journals,* published in 1968, and he suggested that John Wiley and Sons find another author to write about technical and business magazine articles.

I thank my friend, an engineer-writer who is named near the end of Chapter 5, and I thank many other people, one of whom I will name here. He is John Dubas, head of photographic services in Dravo Corporation and an outstanding industrial photographer. He made the photos in Chapter 9. Beyond are teachers, employers, fellow workers, editorial contacts (authors and information sources), and public relations and advertising men. All these people contributed to the 30-odd years of learning upon which this book is primarily based.

Appreciation also is due to some 50 magazine editors who furnished a variety of contemporary material on their practices and accomplishments, personalities, audiences, and relationships with authors in their fields. Drawing from this pool of information, I was able to confirm or modify concepts developed by my own experience in electric power, transportation, and civil engineering journalism.

ROBERT H. DODDS

*Port Washington, New York*
*August, 1969*

# CONTENTS

# WRITING FOR TECHNICAL AND BUSINESS MAGAZINES

# 1

# INTRODUCTION

Two significant classes of writers write for technical and specialized business magazines: editorial staff members, called professionals for the purposes of this chapter, and practitioners of the professions served—here called amateurs.

For amateur and professional alike, the most rewarding magazine-writing product is the feature article, so let's discuss articles.

## WHAT AN ARTICLE IS

Definitions of "article" vary because it is a loose term, but the following are representative.

An article is a story about a specific problem and the practical application of its solution . . . a detailed account of a successful idea having wide possibilities for adoption. . . . An article also may present the lessons of a failure, or it may amplify the general knowledge of the reader.

### To You, the Reader

You, the practitioner in your field, are accustomed to viewing articles from the standpoint of the reader. Your attitude toward articles is essentially friendly, because you are intellectually at home with the magazines that serve your field. When you open the latest issue you know you'll find shop talk: material that will stimulate your work-related thinking and broaden your viewpoint. You know, too, that you can profitably adopt ideas that show up from time to time in the articles you read.

### To You, the Writer

When you're writing an article, you regard it in a different light. Friendship disappears and something of a love-hate attitude develops toward the work. Subconscious or conscious thoughts of school themes and papers (the ones on assigned subjects) erect antipathy barriers, which you struggle to destroy with reason, with fondness for your subject, and with visions of accomplishment and acclaim. The struggle may persist throughout the writing and editing stage, but sooner or later the good guys will prevail, the psychological block will vanish, and you will be able once again to see the article—this time *your* article—as a friend.

## WHAT IT'S LIKE TO WRITE AN ARTICLE

Like a teen-age boy encountering a fact of maturing life, the first-time writer of an article can benefit by the assurance that his experience is common, not unique.

### For an Amateur—the Blank-Paper Block

When an amateur commences an article-writing task he lays a blank piece of paper before him, resolutely grasps his writing instrument—and freezes. Words, phrases, and organizational patterns that were masterfully thought out desert him completely. He clears his throat, shifts in his chair, gets up and goes for a drink of water, knocks the ashes out of his pipe. He stares back at the paper, then averts his gaze, peering here and there in vague, vain search for the missing muse. If there's an external diversion—visitors, a household chore, or a phone call—he secretly welcomes it, insincerely protesting that he's got to write this thing.

Eventually, however, all diversions fail him. Just as a pause for thought can momentarily hush a noisy gathering, circumstances at last will leave the author with nothing to do but put the first word on that blank piece of paper.

Unfortunately, the first word is only the first hurdle. More words must follow. It's helpful to believe that ability to write is mostly willingness to write, but at a time like this he can be forgiven for suspecting that ability transcends both willingness and his own capability.

Thanks to willingness, however, he does achieve a succession of words and phrases appropriate to a general plan, outline, or first draft. Probably he will develop momentum, too, from time to time. Professor Tichy says, "Writing should be kept at a boil." [1] I agree, but reflect that part of what

[1] H. J. Tichy, *Effective Writing for Engineers, Managers, Scientists* (New York: Wiley, 1966), p. 19.

boils during a burst of inspiration is the cold sweat of the previous pause, which was caused by the block of yet another blank sheet of paper.

### For a Professional

When a professional commences an article-writing task he lays a blank piece of paper before him, resolutely grasps his writing instrument—and freezes. Doubtless there are facile exceptions to this generalization, but the whole process of compulsive diversion, dogged application, bursts of progress, and baffled pauses applies to most professionals the same way it does to amateurs. Differences, if any, lie in the relative effects of the respective phases. The initial and subsequent blank-paper blocks may be of shorter duration, thanks to the professional's familiarity with the phenomenon (he knows persistence pays), and his periods of sustained writing (boiling along) may be longer.

These minor differences notwithstanding, it is possible to assert that in the essential act of article writing, amateurs resemble pros remarkably.

How, then, do they differ? The two classes of writers differ significantly in just about every aspect of the technical and business communication process *but* writing. These other aspects form the environment in which the article is conceived, nourished, and published. Professionals live in this environment and they know and understand it in all its aspects.

This mastery is rarely possessed by amateurs, however useful it might be to them. Amateurs are exposed to the environment's aspects in various degrees, and, to the extent that exposure conditions them, amateurs acquire familiarity with the environment.

## THE ARTICLE'S ENVIRONMENT

Unquestionably, the greater his knowledge of the environment, the better the amateur's ability to prepare his article. He'll get to work faster, make fewer false starts, and get published sooner.

This book's mission is to help you do a better job of writing by improving your knowledge of the environment. Writing skill, per se, is within our scope in the manner that the hole is within the scope of the doughnut. Even this introductory chapter is written *around* writing. The ensuing chapters carry forward the same spirit; the only basic assumption is that your skill with sentence structure and syntax is adequate to the writing tasks.

The next five chapters set the stage by discussing motivation, subject choice, and a three-zone region of receptivity. Two assumptions support the motivation theme: first, that you want to get ahead; and second, that you will write about your own field of practice.

The extent of opportunities for writing about your field may surprise you. Chapter 3, "Your Subject," presents both obvious and obscure article types; all are types that the experienced editor uses to assure adequate coverage.

Familiarity with the variety of article subject types is a first step toward comprehensive familiarity with the environment. The zones of receptivity —audience, editor, and publication—round out the picture.

Audience, that is, the readership, requires first consideration because writing must aim at the people to whom the magazine is mailed. Chapter 4, "Your Audience," tells how to get acquainted with the target.

Chapter 5 examines in more detail a principal instrument of audience interpretation, the magazine editor. He is the professional whose business it is to know both your field and your (and his) reader. This chapter deliberately sets out to acquaint you with him as a person and to impress you with his functions.

The remaining stage-setting chapter, "The Publication," has a similar purpose. Magazines play important roles in the fields they serve. These roles, the philosophies and functions of publishing, did not spring up overnight. Rather they matured out of gropings, trial and error, competition, cooperation, and generous helpings of genius during the first half to two-thirds of the past century. Through biography of the technical and business magazine form, Chapter 6 enriches the scene.

Action begins in Chapter 7, which contains advice on conservation of energy (presell your article); then the text proceeds to the tasks of writing and illustrating your article. These two acts star you as the sole performer, and they constitute the core of this book. Covered in Chapters 8 and 9, writing and illustrating logically precede "Editing Practice." This action brings an editor onstage, but you, too, play an essential role in the ultimate steps to publication. Thus Chapter 10 carries the action of "writing for . . . magazines" to its physical conclusion.

However, two more environmental aspects of the overall subject belong in this book. One is the widely practiced catalytic function, public relations, discussed in Chapter 11 as it affects article writing. Chapter 12 is a peroration called "After Publication, What?"

# 2

# WHY WRITE?

A hoary saying that will fetch at least a smile and often outright laughter is, "Us engineers don't need no English."

Really, it's a fascinating gag, combining incongruity with plausibility in just nine syllables. Trouble is, it's neither right nor so.

The truth is, *we need English.* The following incident demonstrates the point.

At the annual convention dinner-dance of one of our engineering founder societies a good many years ago I became intrigued by a happy fellow who was only somewhat older than I. Brashly, I asked, "What do *you* have to celebrate tonight?"

He drew himself up straight, smiled blissfully, and said, "I have *two* things to celebrate tonight: my 33rd birthday and my appointment to full professor rank."

"How on earth did you make full professor at 33?" I asked.

"I wrote!" was his reply.

## PUBLISH AND FLOURISH

This was in the days before the baleful aphorism, "Publish or perish," gained its hold on the teaching profession, so I was able to analyze uncynically the professor's formula for success.

The result of this analysis was a derivation that I used in lectures before engineering students in the late 1940s. In the years since then the instrinsic value of writing as a route to renown has not changed. Witness the words of W. H. Wisely, Executive Secretary of the American Society of Civil Engineers. In an article, "Publish and Flourish," he concludes:

"It is no accident that those whose published work is valued and respected have invariably found success and eminence in their careers." [1]

[1] W. H. Wisely, *Civil Engineering* (June 1968), 29.

The promise of success is a creditable long-range motive for writing, but to achieve an appreciation of the process it is necessary to ask, "What about preliminary and intermediate incentives along the trail to this glowing goal?" Even at that dinner-dance I assumed that the happy professor had no intention of implying that writing is either a shortcut to success or a substitute for hard work; writing is hard work.

By a series of questions I was able to track the formula to its source, and thus to develop a rational derivation. The questions are:

What forms of success are there?
Which forms of success would be enhanced by writing?
What is the mechanism of translating writing into success?
Is there an obligation to write?
What does a person write about?
How did he get to the point of having something to write about?

## EDUCATION STARTS THE PROCESS

Taking the last question first, he got to the point of having something to write about by making, or having made for him, a series of choices. This series began about the time he entered high school, when it became necessary for him to decide whether to take a college-preparatory course. Measurements of his general aptitude and intelligence level helped in this decision. Differences among aptitudes were of further assistance, possibly shaping his high-school program in more detail.

In the latter part of his high-school career, another round of tests helped him discover that he was a prime candidate for technical higher education.

Assuming that he followed this lead and enrolled in a college curriculum such as chemical engineering, physics, business, home economics, or medicine, he eventually became a doer as well as a learner.

The moment he became a doer he arrived at the threshold of having something to write about.

## THE DOER WRITES ABOUT WHAT HE DOES

And what does he write about? He writes about whatever he does that is worth writing about.

At first his employer's opinion will prevail on a subject's value. The boss will expect written reports on special studies, operational procedures, and workaday incidents. He will request drafts for his correspondence, and he will approve written memoranda to other departments. Eventually, as a part of the process of professional and personal maturation, the doer should become the best judge of his work's literary potential, whether it be for internal communication, client service, or publication.

Potential for publication depends on the work's usefulness to the field. Is it a new development that others in the field could apply profitably? Does it broaden or deepen the fund of knowledge in the field? Will it impress, entertain, or otherwise benefit any publication's readers?

The mature professional knows the answers to these questions because he knows the literature of his field. In fact, keeping abreast of the knowledge in his field (the life-long learning process) is an essential element in his ability to do things that are worth writing about.

## AND THEREBY HE PAYS A DEBT

Is there an obligation to write about these things that are worth writing about? I say there is, and support for this point comes readily. E. M. Geverd, Administrator of Staff Publications, Radio Corporation of America writes:

"Every engineer has the responsibility for promoting a better understanding of his work and his profession and for broadening public and professional knowledge by communicating what he has learned." [2]

Moreover, the obligation exists regardless of the possibility or probability of future personal reward to the writer. Rather, it is to the past that the obligation is owed.

In large measure, a professional's proficiencies are acquired through the medium of literature—books and periodical publications. In his school days, when he is learning fundamental principles and basic facts, he depends mostly on books. Academic officers direct his attention to the appropriate literature and help him understand it. Ordinarily they supplement assigned-literature data with material they have gleaned from other literary sources (or from their own working experience, including research). As the professional matures, he keeps up to date by reading the periodicals of his field, by hearing papers presented at his society meetings, and, when he decides it's necessary, by studying appropriate books. Throughout the learning process, therefore, the professional receives information from the literature of his field. This information, or at least some significant portion of it, is essential to his performance, and hence to his livelihood.

[2] E. M. Geverd, *American Engineer* (now *Professional Engineer*) (November 1967), 39.

Thus it may be said that the professional receives a valuable consideration from the literature. He receives word of the innovative thoughts and manipulations of thousands of his predecessors and peers; he benefits from the integrative skills of writers, editors, and professors who weave fact and principle into systematic discipline and subdiscipline.

In short, he incurs a debt to the literature. There is little that he can do to repay this debt as long as his work duplicates the work of others. But it is not the nature of the professional's work to remain static. It demands innovation—new combinations and applications of existing art; original things, bigger things, and tinier things to accomplish civilization's ends; ideas that are timely for changing economic, social, and political conditions. By responding to the demand for useful innovation the professional creates the capability to repay in kind his debt to literature. He writes about his innovation, making it available to thousands of his peers and successors.

This act of writing is commonly referred to as "contributing to the literature." If you agree that a creditor-debtor relationship exists between the literature and the learner, you also will agree that the act of writing can be called "repaying the literature."

By repaying the literature, you consummate the second step of a fair trade that can be summarized as follows.

1. You receive most of your education and help in your experience from the fund of professional literature.
2. When you develop new knowledge and you contribute to the fund of literature, you reimburse the fund for its contribution to your accomplishment.

## YOUR PROFIT: RECOGNITION

So much for the fair trade. Now you're even. How about the profit? Recalling our list of questions tracing success through writing back to the beginning of learning, the next question is, "What is the mechanism of translating writing into success?"

The basic achievement of successful writing is *recognition*. We are most concerned with the recognition that comes from publication, particularly in business and technical magazines, but the process of "write and be recognized" takes effect in a limited way even in the area of private communication. Good business letters impress both the addressee and your boss. Next, at the borderline of public communication, is the report. Good reports impress decision makers and at least the few or many who are affected by the decisions. Really wide recognition, however, is accorded to the writer of magazine articles, journal papers, and books, for the simple reason that publication permits mass exposure of his work.

When you publish, readers identify you with the subject matter of your writing, and they tend to credit you with superior knowledge. Thus you become known as an expert in the subject of your writing, whether it be the water-cement ratio in concrete mix design or the whole field of computer software. James M. Lufkin, manager of professional publications for Honeywell, Inc., writes: "The professional man who has published two or three articles on his subject in respected journals is almost by definition an authority." [3]

Furthermore, you become known by people who can do you good: people who read your writing because they believe they need it to solve specific problems, or to keep up with the fast-moving pace of technological advances in your field; people who might need you as a consultant or name you to an important committee.

Implicit in the use of the word "recognition" is that the recognition is favorable. Actually, there's not much likelihood that it would be unfavorable. Editors are experienced and astute judges of their fields, and they have their publications' reputations to think of. Poor material reflects unfavorably on author and publisher alike, so the editor can be counted upon to reject ill-conceived contributions.

This is not to say that the editor will protect you from criticism or controversy. Lots of technical and business writing provokes disagreement; some of it is as valid, sincere, and well-informed as the writing it attacks. But honest controversy neither impugns the truth-seeking merits of conflicting positions nor cheapens the recognition gained by the participants. As a matter of fact, published discussion of your work by your fellow professionals earns more attention and recognition for your work. Discussion also benefits the publication by spicing its basic fare of valuable information and thereby heightening reader interest.

Reader interest in your writing accords recognition to a secondary beneficiary—your employer. He basks in your reflected glory, sharing your recognition and reaping rewards that are analogous to yours even though they may be less tangible.

## THE FRUITS OF RECOGNITION

And what are your rewards? Which forms of success may be enhanced by writing? Recognition itself is a form of success: the favorable opinion of

[3] J. M. Lufkin, *American Engineer* (July 1968), 25.

your contemporaries; professional prestige. Self-satisfaction is another form: satisfaction at having paid a debt to the literature; the simple pleasure of completing a writing task; the joy of seeing your work in print. The opportunity to do still more interesting, significant, and original work is a real incentive, or form of success, for many. Recognition of the writer's capabilities for significant work brings new assignments that he might not otherwise receive, whether they be in his own department, in another part of the organization, or with a new employer. Recognition also may bring consulting assignments, usually involving challenging technical and/or political problems.

Consulting firms, particularly those in fields that have inhibitive rules about advertising, deliberately seek recognition through writing as one means of business promotion. Commercial and industrial concerns release technical articles and reprints as a facet of their overall public relations and advertising efforts.

These assertions about company activities may appear to stray from the subject of why you should write, but you remain the individual source of written work. Because of the near identity of purpose held by the employer and the writer, companies offer various kinds of encouragement. Examples are time for writing, secretarial help, illustration services, and monetary awards.

Money, of course, is honored by many people as a form of success. It, too, can come to you through recognition generated by your writing. Money awards by employers and money honoraria from publications are titillating. Royalties from books may be equally pleasant (they might be substantial). But the really gratifying financial reward to the person who writes about his work is likely to be found in his regular pay envelope.

At the very least, publication will be recognized as an element of satisfactory job progress, and, hence, of eligibility for merit increases in salary. More often, writing will be considered evidence of superior performance, bringing better raises, incentive compensation, or bonuses, and enhancing the chances of promotion.

There's the possibility, too, that your writing will attract the attention of another employer. Or you may be seeking a change. Mr. Lufkin states:

". . . Of two résumés, describing two engineers of similar ability and professional experience, the one that concludes with even a short list of publications has an undeniable advantage in the eyes of most employers." [4]

Whatever the scope of incentives offered by a new employer, the package is almost certain to contain more money.

[4] *Ibid.*, p. 25.

The increased possibility of losing you is a risk your present employer takes when he encourages you to write. The mutual advantages (identity of purpose) are so preponderant, however, that the intelligent employer does not hesitate to "expose" you.

Exposing your expertise can bring yet another form of success, a form that for many people is sweeter than all else—influence. In addition to the influence of your writing itself there is a large area of consequent influence available—committee work and office-holding in significant organizations. Just about any work-related subject that's worth writing about also is worth having a committee about. These committees are found in technical societies and trade associations, and they concern themselves not only with internal organizational matters but also with external professional-community affairs like laws, codes, standards, ethics, and awards.

For example, the professional engineers society of a state, through one of its committees, advises the legislature on the professional engineers licensing law. Committees of the American Society of Mechanical Engineers develop technical standards that are accepted internationally. (In turn, writing is the usual instrumentality of committee influence.) Thus it is that committees have various kinds of influence on technical and trade affairs as well as on society as a whole or in part.

Your writings are quite likely to be seen by committeemen in your field, because they have a responsibility to keep up with the literature. It also is quite likely that committeemen in your field are looking for new blood—for people who will work. By your writing you have demonstrated a capacity for work.

All you have to do is say, "Yes." From there, assuming continued hard work, it's but a step to the chairmanship, then membership on the next committee up the ladder, and so on to ever greater, more gratifying influentiality.

Of course there are other writing-related approaches to achieving some form of influence (such as better jobs for more influence on fellow workers) and there are other ways of getting "discovered" for committee work (recommendation by fellow workers, etc.), but for the person who strains against the bonds of anonymity, writing and consequent committee work are readily available, one-two strokes to renown.

These, then, are forms of success that may be enhanced by writing: influence, money, more interesting work, self-satisfaction, and, generally, the recognition of your worth by your fellow professionals.

The next question is now anticlimactic in the context of this book: "What forms of success are there?"

The question becomes, "What *other* forms of success are there?" There are many, of course: success in family life; successful adjustment to matters of religion, conscience, and social obligation; satisfaction with a hobby or avocation; achievement of pleasures and happiness. The chances are, however, at least for the ambitious, work-oriented professional, that these achievements of other forms of success are concomitant with, if not dependent on, the kinds of success asserted here to be possible through writing about one's work.

## WRITING IS AN ESSENTIAL WORKING TOOL

In short, the pen is a useful instrument for achieving success; it also is an essential working tool. This is the final point to be made in answering the question, "Why write?"

To substantiate the point that the pen is an essential working tool, it is necessary to refute without humor the statement, "Us engineers don't need no English." For the sake of this argument, the term "engineers" can be taken to mean any of the professional persuasions to which the lines of this chapter apply. Each of the professions has its lore, its jargon, its tools, and its trappings; and each has its interface with the world it serves.

Engineers have their educational disciplines, their special vocabularies and symbols, and their instruments, logics, machines, slide rules, and license certificates. They also have a very broad interface with the nonengineering world of people who buy their products and services, vote for, and finance their projects. Except for starry-eyed buyers of automobiles, few people accept the engineer's bounty on faith alone. In fact, some good things the engineer "sells" just naturally generate sales resistance. For example, sewage treatment is a distasteful subject and a sewage treatment plant costs money. It's difficult to generate much enthusiasm for something that's both expensive and distasteful. Yet pollution control is an essential element of public activity, and the eventual result of discharging raw sewage into a stream is more distasteful and probably more expensive than building the facility in the first place.

The engineer determines the needs for antipollution works (and for hundreds of other engineered things), and then interprets them to the appropriate "public." Hopefully he interprets his work in such a way that the public will understand and agree with his determinations.

Now here's the point. To do his whole job an engineer must translate his work from engineering into English. This "must" implies equal proficiency in engineering and English, and it applies to every engineer who would merit the title.

The need to communicate effectively is most apparent at the engineer-public interface, but it pervades the practice, too. The classic engineering method is to divide and subdivide a problem into manageable components, then synthesize the subsolutions to achieve over-all solution. The solution of all but the simplest engineering problems is an administrative-organization process comprising two or more levels of responsibility. Dividing the problem requires effective communication from the top downward to the component solvers; synthesis requires communication upward from those solvers. Written communication both downward and upward saves time ("How's that again?") if its English is lucid. Lucid or not, however, writing is essential to engineering, and anyone who says he "don't need no English" is either wrong or not an engineer.

Why write?

*Because* writing is part of your work.

*Because* you owe to the literature an account of the new knowledge you develop.

*Because* recognition will surely result from your contributions to the literature.

*Because* recognition will help you achieve many forms of success.

And the beauty of it all is that writing is a tool that you control yourself. You can decide what to write and when to write. You also can master the process of placing your writing in the right publication.

# 3

# YOUR SUBJECT

The person who usually contributes to technical and specialized business magazines does not have to search for a subject. His subject is self-evident in his work—or so it would seem.

Actually, many a worthwhile topic remains unexplored because its potential amanuensis fails to consider its article possibilities. Still more subjects suffer from the mistaken belief that they're not worth writing about.

These deficiencies would disappear if every developer of new knowledge were familiar with the variety of subject types that make good copy. These subject types are discussed in this chapter. Hopefully the discussion will send the reader at last to the Dictaphone or typewriter to tell the world some useful things he knows.

The measure of this chapter's value, then, will be the extent to which it stimulates articles that otherwise never would appear.

## SOURCES

Your work, of course, is the normal source of subject matter for your technical and business magazine articles. Essentially, "your work" means your means of gainful employment, but your professional society and trade association activities also can yield material that will expand the body of literature beneficially. Moreover, you may encounter the opportunity or obligation to report the work of others in your field.

*Subjects may be classified approximately as fact and opinion.* The classifications are "approximate" because opinion requires facts for support and factual presentations necessarily involve the writers' and editors' opinions of the facts' relative importance. In addition, between mostly-fact and mostly-opinion there lies a middle ground of theory and speculation that is more nearly a 50-50 blend.

## FACTS PREDOMINATE

Fact is by far the more important element, and it is the one in which the creators of new knowledge come nearest to receiving an even break from both editors and readers. The information itself commands attention; the author's renown or lack of it will have little bearing on acceptance.

Types of fact articles are the following:

- New developments
- Superlatives of existing art
- Major plans
- Progress on significant developments
- Unusual (or bizarre) subjects
- State-of-the-art
- Stories worth repeating
- Proof of the clever idea
- Failures

The nearest thing to a common denominator among these types of articles is *newness,* just as the most important reason for subscribing to technical and business magazines is to keep up with what's new in the field. The periodical form of publication naturally satisfies this objective. ("Stories worth repeating" would appear to be our exception to the common denominator of newness, but these stories would not be repeated unless the editors decided they would be new to a significant segment of the readers.)

### New Developments

In a sense, under the common-denominator concept, all subject types mentioned above are new developments, so what we're writing about here should be circumscribed. These new developments are the inventions (patented and unpatented), the improved designs, products, and procedures that you, the developer and hence the prospective author, would be most likely to think of when you think of subjects for articles. These are developments that probably are entirely within your experience and in which you have a continuing interest, so they are easy to write about. Required research presumably was done during the development process. Practically everything you need is at your fingertips: personal knowledge that is waiting to become written words through the medium of pen, pencil, or typewriter; illustrations, already in hand, that may require no more than editorial attention to ready them for the press.

### Superlatives of Existing Art

Most subjects of this type take a little imagination and research to demonstrate their record-breaking character and hence their newness. Some, however, are fully as obvious to their progenitors as are the neatly scoped new developments categorized above. New York's Verrazano Narrows Bridge was sure to have the world's longest main suspension span as soon as the plans were adopted and the money committed. Little was involved in the way of design or construction innovations; yet the project was news throughout, and it continues to command attention.

The World Trade Center towers, twin examples of superlatives, embody the added attraction of significant innovation. These towers will soar at full-floor section some 100 feet higher than the tapering tiers of their uptown neighbor, the Empire State Building. To resist unprecedented wind loadings the structural engineers designed each tall wall as a truss, thus creating boxed vertical cantilevers. (Chicago's John Hancock Center is another example of this approach.) Predesign studies included exhaustive wind-tunnel tests at Colorado State University, and the findings are available as a matter of public record.

Probably each of the succession of "smallest" hearing aids was not difficult to credit, either. The manufacturers presumably knew their competitors' products and were able to make their claims accordingly. Whether they did it accurately or not could be an open question. George Rostky, Editor of *EEE,* writes on the importance of checking claims:

"When a company introduced the 'world's smallest capacitor,' *EEE* checked the dimension and researched its files. It reported that the new 'world's smallest' was 'almost as small' as a unit introduced by the same company a year earlier." [1]

Many record-breaking performances, however, are accomplished without thought of their superlatives. To the engineer, the project is just another highway (or ammonia plant, or data-processing system) with the usual special problems. The special problems were investigated to a sufficient extent to satisfy the engineer that the solutions were sound, but there was no time to pause and ponder publishing possibilities.

Later, when an editor or public relations man asked whether any aspects of the project gave it publishable significance, the engineer scratched his head and started to check over the "usual" special problems to see whether any of them were, in fact, unusual. He may have come to suspect that he

[1] G. Rostky, "A Technical Magazine Exciting?" Unpaginated reprint from *Better Editing* (Fall 1967).

had designed the lowest cost-per-kilowatt power plant of its size or the highest-capacity code system for digital data transmission in a railway signaling system. (Both these superlatives could be incidental to unique project conditions for which problem solutions were well within existing art.)

Having isolated a subject by using his imagination, the prospective author now needs to find out whether he does, in fact, have a first. Research is the answer, the extra step that can change an easy writing task into a time consumer. A literature search is likely to be fruitless. Data on the less obvious superlatives have a way of showing up in the literature as letters to the editor after someone has published a hopeful claim. Manufacturers of any equipment incorporated in the project can be helpful, but their competitors should be checked, too. Fortunately, technical and business editors not only like superlatives, they also are widely informed and delicately tuned on these reader-catching themes. Accordingly, the author may be able to save himself a lot of research time by getting in touch with the editor of the magazine he has in mind for publishing his article.

## Major Plans

Plans for major projects may be news, depending mostly on one or both of two plausibility premises. The more influential premise is the prospect (or, better, the commitment) of money for the work. The second is the relative financial or technical prestige of the person or agency announcing the plans. A promoter with a record of successful accomplishment can command attention for any scheme to which he is willing to put his name. A famous technical authority or an important governmental construction agency, too, will be heard.

Technical concepts and details are likely to be sparse when plans are preliminary, so the prestige of the source must substitute for proof of validity by facts and logic. Later in the process, when the details are worked out—when plans and specifications are ready for bidding, for example—the person who worked out the details will be in a good and acceptable position to write about the project.

Beyond straightforward project description, an article concerned with major plans should tell about important design decisions (steel versus concrete; microwave versus wire) and the analyses that led to them. Cost usually is a significant element of articles at this stage. Planned innovations, minor in the overall project picture but potentially good copy, may be mentioned, but prudence dictates a tone of "wait and see" as far as performance claims are concerned. The type of article discussed here commands attention because of the size of the project discussed, not the cleverness of its details.

### Progress on Significant Developments

This subject type is what newspapermen call a follow-up. It applies to subjects that have been reported previously as either major plans or completed innovations.

Just as readers want to know what important things are being planned, they want to know how these things are getting along. A status report—construction in place; components transported to assembly point—lays the groundwork for the story. Details of methods, materials, and organization can be presented as accomplished facts. Design theories can be illustrated, and the procedures planned for the remainder of the work can be revealed.

The expression "progress on completed innovations" may seem to involve a contradiction, but again, it's the reader's need for follow-up data that the magazines seek to satisfy. For example, as first described in an article, the new synthetic fiber plant operated at rated capacity, its output being of standard quality; then performance fell off. To what extent did it fall off and why? What steps, if any, were taken to correct the situation? Another example: traffic across the toll bridge exceeded forecasts by 8–11% in the first 5 years of operation. How does this deviation alter the schedule for expansion of the facility? Does this experience yield lessons that improve the science of traffic engineering? Both failure and spectacular success expand the fund of knowledge, and they create not only the opportunity to write but the obligation.

### Unusual Subjects

Like the term "new," the term "unusual" applies to topics in periodical literature. Here again, a term that can have broader meaning is intended in a restrictive sense. We do not include the useful new development or the superlative; the former is expected to become usual; the latter has lesser prototypes, and it probably will be bested.

This "unusual," like the human it imitates, is the incongruous, the familiar subject in strange surroundings, or vice versa: the bizarre solution of an ordinary problem (an old lady's house in the form of a shoe).

Your clue to the possibility that you have a story of this type is a fellow worker's remark, "Hey, that's a funny house you're building." Maybe the editors will agree and the subscribers will pause to read your article. Even technical magazines serve appetizers.

## State-of-the-Art

Quite the opposite position on the menu is held by the state-of-the-art piece (some call it a review; others, a "tutorial" article). Truly a *pièce de résistance,* it is more an essay or a monograph than an article. It is included under the general heading of "your work," because the acceptable contributor of state-of-the-art material just about has to be an expert in the subject about which he writes—not necessarily renowned, but an expert nevertheless. To be sure, most of his piece will be about the work of others, but it is his research and his knowledge of the subject that make the assembly cogent.

This is not to say that the overwhelming bulk of state-of-the-art is authored by experts in the field. Far from it. Perhaps because many publications insist on maintaining scheduled series of these features, and perhaps because their preparation is such hard work for people who have their regular jobs to do, state-of-the-art writing is widely practiced by editorial staffers. Afer all, it is their regular job to carry out the publication's policies, and they can neither wait for casual submissions nor count on assigned contributions to meet the schedules. It should be added that many of the editors who write state-of-the-art pieces are authentic experts in their fields.

## Stories Worth Repeating

Familiarity with subject matter sometimes lulls a potential author into the mistaken opinion that everybody knows about it—particularly if "it" is already in the literature. If you are the potential author, ask yourself two test questions, lest you lose a good subject by default:

"*Does* everybody know about it?"
"Would it interest or be useful to those who don't know about it?"

The chances are highly favorable that for all but the most overworked topics the answer will be "no" to the first question and "yes" to the second.

A classic example of a story that has borne repeating is that of parallel-wire cable spinning for a suspension bridge. From the Roeblings' Brooklyn Bridge to O. H. Ammann's Verrazano Narrows Bridge the method remained essentially the same, and despite developments of bridge strands and of prebundled parallel wires, on-site spinning endures as an important method and standard of comparison.

The big suspension-bridge jobs occur so infrequently that each one becomes a training ground for a new generation of bridge builders—and construction fans. It is a safe bet that the cable spinning for every project that utilized the method has been described somewhere in civil engineering literature. (It is an even safer bet that the few really long cables that were placed by other methods were written up.)

Another story that's worth repeating is the one that was ahead of its time when it first ran. In the mid-1930s Sergey Steuermann wrote for *Engineering News-Record* on his new vibroflotation method of consolidating loose sand. The method had been proved by full-scale experiments in Europe. The Editor, F. E. Schmitt, told Mr. Steuermann it would take 10–15 years for the idea to catch on. Five years after World War II Mr. Steuermann got his first commercial vibroflotation job, and upon its successful completion *Engineering News-Record* published (repeated) the basic story.

Most repeatable stories are essentially "how-to" articles. Many an author, expert and free lance alike, has mined this vast lode and found markets for his metal. Usually there's at least one magazine in a field that's ready for the repeat of a "how-to" article.

The story worth repeating may be characterized further as the story of a contemporary, still up-to-date activity. The who, where, and when will be different, but the what and probably the why will be the same.

## Proof of the Clever Idea

Here is another story category that, like repeatable stories, tends to be underplayed. Stories that should be repeated, but aren't, suffer from the potential authors' deprecative thoughts about their material's value. Many possibilities for proof-of-idea articles lose out because their authorities don't think about the material at all. Moreover, even if they do consider the material's journalistic possibilities, many of these people take no action. They're too busy solving newer problems to pause and report on the recent past.

This forward-looking attitude is good for local progress, but it's frustrating for editors and unfair to readers. When the bright ideas are hatching, their authors are eager to talk about them. The editor may give some space to the ideas, but he'll hold off the full treatment until results can be included. Then, even though the results prove out the ideas, it's like pulling elephants' teeth to extract the whole story from the now-diverted author.

We can lament the author's missed opportunity for recognition and sympathize with the editor's frustration, but the reader stands as the real loser because he is the would-be beneficiary of the presumably useful experience.

So think of the reader and your obligation to the literature and the opportunity for recognition every time your ideas turn out well. You may have first-rate stories to tell.

It should be added that there is one circumstance, not an uncommon one today, in which the proof of the bright idea will receive ample consideration: When a public relations man is active in the organization. Once he gets wind of an idea—and he's likely to learn about it at the talking stage—he'll follow it as a publicity possibility as long as there's a line of favorable copy in it.

## Failures, Too

Unfortunately, not all ideas turn out well, and therefore some follow-ups would be considered unfavorable copy—at least by public relations men and most other people in the organization responsible for trying out the big idea. The profession as a whole, however, is eager for failure stories, not because individuals and organizations wish to gloat over their contemporaries' misfortunes, but because they constantly strive to learn more about the limitations of their art.

A successful idea or problem solution may be well within the bounds of safety and practicality; as such it will inspire adoption, adaptation, and imitation. A failure accomplished by honest, intelligent, educated, experienced men can extend a frontier, halt an unsuspected dangerous practice, or open up a new field.

Aerodynamic stability of suspension bridges was an unknown (or forgotten) subject in the late 1930s, when plans were drawn for the Tacoma Narrows Bridge. Most previous suspension bridges were heavy, stiff, and broad relative to their length. They had low "slenderness ratios," to borrow a term from Tacoma Narrows Bridge designer Leon Moisieff, who wrote early in 1940 on the trend to higher slenderness ratios and on the fact that his new bridge had the highest ratio of all.

Soon thereafter the bridge developed alarming motion during high breezes. Wind-tunnel studies of a model were begun at the University of Washington. Corrective measures developed from the tests came too late to save "Galloping Gertie"—she collapsed on November 7, 1940—but the results were directly applied to stabilize other slender suspension spans, New York's Bronx-Whitestone Bridge among them, and aerodynamic analysis became an accepted element of engineering for important structures of various types.

These widespread benefits from the Tacoma Narrows Bridge failure were no accident. Engineers worked hard to lay bare new truths discoverable in the incident; engineering magazines eagerly published discussions, conclusions, and opinions, to the end that everyone with an interest in testing methods, materials, dynamic forces, structural analysis and design, bridges, and construction had ample opportunity to get the full story.

Of course, human nature being what it is, not all failure stories are as highly visible and easy to get as the Tacoma Narrows Bridge story. You yourself may be tempted to cover up or forget a flop you've fostered. But before you do, think well on the possible benefits of disclosure to your field. If the experience should be told, swallow your pride and write the facts.

## MAGAZINES WANT YOUR OPINIONS, TOO

Being at the opposite end of the scale from fact, opinion might be expected to be difficult to sell. Actually, it's not all that discouraging. If your topic is of general importance, your logic is plausible, and your supporting data are new, you are in a good position. It matters not that you are "unknown," although a visible reputation for being right won't hurt your mission. An open-forum spirit, found in most publications, sustains this happy circumstance.

Most writings of opinion take the form of letters to the editor, and they usually involve comment on matters that have been covered previously in the magazines. Prior coverage is by no means a prerequisite of reader comment, however. Anything that is topical in the field of the magazine's coverage is fair game.

The guest editorial is a related form of opinion presentation. The editors may dignify an outstanding and timely expression of opinion by giving it special position. A few publications make a regular practice of printing guest editorials by recognized experts, but for most periodicals these features are matters of opportunity and special purpose. And the editors are more likely to invite them than wait for them to come in "over the transom."

### Controversy Stimulates Readership

Another class of material that belongs under the heading "opinion" is the controversial article.

Among technical magazines there is a trace of tradition against fostering controversy ("Let's keep peace in the family"), but most are pleased to publish controversial articles as long as they meet some criterion of dignity. Many magazines pursue controversy energetically as a device for drawing readers' attention to important topics.

Acceptable topics do not involve personalities. Rather, they treat bona fide alternatives within the fields served by the magazines.

In its "Author's Guide," *Industrial Research* lists "Articles on the major controversies of science and technology" as one of four principal categories sought from contributors. Typical titles given as examples are the following.

"The Case for Going to the Moon."
"Unionism in the Laboratory."
"Liquid versus Solid Fuels."
"Mohole: Boondoggle or Bonanza?"

*EEE* Editor Rostky deliberately sets up a "Product Battleground" in his pages, offering equal space to competing types of products.

"Forcing competition to slug it out in public instead of in private corridor discussions is a valuable service to readers who must make serious buying decisions. And it makes for exciting reading." [2]

Writing more specifically to the point of controversy, Mr. Rostky states:

"Most [business paper] editors shun controversy. Yet controversy is what life is made of. Controversy is just another word for competition and that's what business is made of. Competition forces companies to advertise and makes people read to learn how to do their jobs better. It sits on almost every business decision." [3]

(A Rostky pearl: "Readers appreciate opinions as long as they can recognize them as such.")

## REPORTING THE WORK OF OTHERS

What this book is mostly about is writing about your own work—that is, work that you perform or supervise. There is a broader field of subject matter about which you are qualified to write by virtue of your own work. You can write about the work of others in your field. Being familiar with your field, you can judge the significance of your fellow professionals' work and you can write about *their* work.

A simple situation that could bring about this action would be a boomerang reaction to the question (by you): "Why don't you write about that?" The person who developed "that" is disinclined to write, so you volunteer to do the job. You place yourself in a quasieditorial staff position, as a volunteer reporter. In other words, you undertake to do what editorial workers frequently do when stories they want aren't forthcoming from the logical authors. You write the story.

Editors welcome many types of reported material. A common reporting

[2] *Ibid.*
[3] *Ibid.*

topic is convention coverage. Editorial staffers may not be available for attendance at a meeting where significant developments might occur. You happen to be going, so you offer to watch and listen (and take notes) for newsworthy revelations in speeches and papers, and you send your story to the editors. (Don't forget to list the new officers, and be sure you make the deadline.)

Other kinds of news of your field are welcome, too: important project announcements, personal items, corporate developments, legal cases and decisions. Many a professional has made a tidy sum on the side by becoming a correspondent (stringer) for the news magazine in his field. It's not necessary, however, to have a regular assignment for your items to be acceptable.

Book reviews are widely welcomed by editors. In fact, if you want a certain new book and you're willing to review it, there probably is a publication that will give it to you—and pay you an honorarium besides.

## KNOWING WHAT WILL SELL

In the foregoing discussion, words like "topical" and "contemporary interest" impose a certain burden of knowledge on an author, over and above his detailed topic knowledge. These words both imply and require a knowledge of informational needs and acceptability.

In fact, "feel" is a better word than "knowledge" in the present context. The concept is a cousin of the journalist's "nose for news," with the important difference that the journalist writes others' news; you write mostly about your own.

Your "feel" for what is topical in your field probably is in pretty good shape already; conscious application will develop it further. You are well informed because you read at least one magazine that keeps you up to date on the field, and therefore on what is of contemporary interest. Study several back issues of this magazine and you will confirm what you already "feel." Now extend your study of topics to other magazines in your field and related fields; soon you'll be nearly as well versed in contemporary interest (and receptivity) as the magazines' editors are themselves.

"Selling" your subject, or your article, is a definite step in the process of publication, and it is treated in Chapter 7, but before you take that step you should assure yourself that you are familiar with three external influences on the acceptability of your subject: the reader, the editor, and the publication. Of the three, the reader (audience) is the most important, and it is the subject of the next chapter.

# 4

# YOUR AUDIENCE

"Your Audience" is a better title for this chapter than "Your Reader," even though the latter term describes the target of your writing more precisely.

Significantly, "audience" permits a performing-arts analogy that dramatizes the importance—and difficulty—of reader visualization. Like the audience at a play, concert, or lecture, your readership responds to your performance. Of course readers rarely applaud and they always can "walk out" without embarrassment, but the qualitative facts of expression and reception exist for both situations.

Missing from the writer-reader relationship, however, is the possibility of instant interplay between performer and witness. The performer can see his audience, size it up, and adjust his actions to his audience's level of receptiveness. Thus he communicates with the aid of feedback. The author has no immediate feedback from his reader; yet he must cater to his reader's needs. Otherwise he may fail to communicate effectively even though his subject is timely and his writing is lucid.

Your ability to visualize your reader—to address him in his language and in terms of his interests—is an indispensable part of your writing equipment. Strive at all times to keep your reader in mind.

## HOW TO KNOW YOUR READER

Fortunately there are good ways of knowing the readers of technical and business magazines.

### Be the Reader

The easiest way to know your reader is to *be* one already. In this happy case you can write purely in terms of your own interests and be assured that you will stimulate maximum readership. The publication you write for will be one that serves your field in general and you in particular. Probably many articles in this publication have helped you in your work—possibly even helped you to create the subject of your article. In this case a good first test of your ability to visualize the reader could be, "Would I expect to read about my subject in this magazine?" If the answer is truly "yes," you know your audience.

However, this is a circumstance in which there will be a tendency to arrive at a "yes" answer a little too soon. Consider the breadth (or narrowness) of your subject. Would your natural approach omit background information that would be required by many of the magazine's readers? Consider the intellectual level of your approach to the subject. Is it too highbrow for them? Your familiarity with the magazine will permit you to arrive at proper appraisals of these questions.

The results of your appraisals will permit further analysis to determine whether you can modify your approach and treatment to suit your pet magazine's needs. If you cannot, you should pick another, more suitable magazine. To find a more suitable magazine, ask fellow workers for suggestions and check through lists such as Appendix C of this book. Most technical and business magazines' titles reflect their scopes.

## Study the Magazine and Its Readers

The right magazine may well be a stranger to you, in which case you will have to depend on more than your intuitive powers for audience understanding. Three sources of useful information on audiences are the magazines themselves, their circulation statements, and their editors. Consideration of two or three of these sources is better than reliance on one.

*Become the Reader.* If you feel insufficiently familar with the people who receive a magazine, become a reader yourself. You may already have inspected the magazine from the standpoint of topic acceptability; now try for the reader's viewpoint. Study several issues; guess the occupational interests of persons who would benefit from each feature: articles, editorials, regular departments, and advertisements. See if you detect useful storytelling techniques. Is the editorial content pitched to a particular level of competence?

Advertisements, too, are reliable indicators of a magazine's audience, because advertisers are hard-nosed about where they spend their money. They want their ads placed before the eyes of the maximum number of buyers and specifiers of their products, and they take great pains to know which publications will fill the bill.

*Study the Circulation Statement.* This document, which most publications issue, is the advertiser's basic source of data for media analysis. He supplements it with readership studies, publishers' statistical manipulations, and anything else he can lay his hands on, but for our purposes the circulation statement is a sufficient tool.

In its simplest form, a magazine's circulation statement presents the number of copies sent to each of several specialty occupations within the field served, and it tabulates the distribution geographically. The statement also describes the field served. Thus the "ABC" statement for *Hydrocarbon Processing* declares, *"Field Served:* Specialized for Oil, Gas and Petrochemical Processing covering Engineers, Operations and Management."

"ABC" stands for Audit Bureau of Circulations, an organization that grew up many years ago in response to a need by advertisers for reliable circulation figures. (Some old-time publishers concealed or inflated their figures; ABC made these practices obsolete.) ABC members developed standard forms for reporting circulation breakdowns, thus enhancing comparability, and they submitted their circulation files to audit, establishing reliability.

Circulation files on ABC member publications disclose, in addition to names and addresses, the occupation of each recipient and whether he has a bona fide paid subscription. Free publications, called controlled-circulation magazines, have similar voluntary associations, which audit substantially comparable data. Instead of claiming numbers of paid subscriptions, however, controlled-circulation magazines certify the recency of "qualification."

Qualification is the key to the circulation statement's usefulness to advertiser and author alike. Both paid and free magazines attempt to restrict their distribution to people within the field served, in order to convince advertisers that their messages will not be wasted on nonbuying influences—or on unwilling recipients, for that matter; there must be some form of positive indication from a recipient that he wants the publication.

Whether or not a subscription order is accompanied by a check (or promise to pay), the order must state at least the sender's business connection and title. Some publications, particularly free ones, ask prospective subscribers to fill out long checklists pertaining to their buying influence. These "bingo cards" become raw data for detailed circulation analyses, many of which go far beyond the breakdown presented in the ordinary circulation statement. The complex analyses become ammunition for advertising space salesmen, but they probably are unnecessarily detailed for a prospective author's use.

Reference to an ordinary circulation statement will permit you to judge its possible value to your process of becoming familiar with a magazine's readership. The following table was taken from the Business Publications Audit of Circulation, Inc. (BPA), "publisher's statement" for *Building Construction.*

**BUILDING CONSTRUCTION** **DECEMBER 1967**

**3a. Business/Occupational Breakdown of Qualified Circulation for Issue of November 1967**

This issue is √ 0.4% or √ 155 copies above average of other 5 √ issues reported in Paragraph 2.

| | | | Classification By Title | | |
|---|---|---|---|---|---|
| Business and Industry | Total Qualified Copies | Percent of Total | Architects | Engineers | Owner, Presidents, Vice Pres., Gen Mgrs., Supts., and Other Personnel Who Specify Bldg. Products, Tools, and Equipment |
| Architectural | 12,760 | 29.3% | 12,080 | 141 | 539 |
| Engineering | 5,020 | 11.5 | 272 | 4,045 | 703 |
| Architectural/ Engineering | 5,248 | 12.0 | 3,499 | 1,279 | 470 |
| Architectural/ Engr./Contracting | 1,461 | 3.4 | 559 | 550 | 352 |
| General Contracting | 13,600 | 31.2 | 518 | 2,068 | 11,014 |
| Sub-contracting | 3,271 | 7.5 | 217 | 852 | 2,202 |
| Government | 1,192 | 2.7 | 673 | 364 | 155 |
| Arch.-Engr. or Design Depts. Commercial, Industrial Firms, and Institutions. | 1,063 | 2.4 | 362 | 429 | 272 |
| Total | 43,615 | 100.0% | 18,180 | 9,728 | 15,707 |
| | | | 41.7% | 22.3% | 36.0% |

In its introduction, the BPA statement form calls for paragraphs on "field served" and "definition of recipient qualification." For *Building Construction,* the paragraphs are the following.

*"Field Served. Building Construction* is a professional magazine serving individuals in firms engaged in the architectural and engineering design,

and construction of industrial, office, commercial, religious, educational, institutional, apartment, hotel, and other similar buildings, exclusive of private homes.

"*Definition of Recipient Qualification.* Qualified recipients are: owners, presidents, vice presidents, other company officers, general managers, managers, and other supervisory personnel; architects and engineers; personnel who specify building products, tools and equipment in Architectural, Architectural/Engineering, Architectural/Engineering/Contracting, Consulting Engineering, general contracting and sub-contracting firms."

*The AIA Data Form.* Many publishers prepare another standardized form, which is more informative in some ways than the circulation statement. This is the Association of Industrial Advertisers "Business Publication Data Form." This form, running 10–20 pages, depending on the amount of information the publisher wants to present, summarizes circulation-statement data (sometimes by reference to the statement) and also includes material on universe, advertising, editorial, and readership.

"Universe" means the field served. Usually there is an attempt to define the size of field with census data on standard industrial classifications available from associations and governmental agencies. Advertising data give numbers of pages of advertising in various classifications. These classifications constitute another clue to publications' readership. The section on editorial states editorial philosophies and fields and subfields covered. Numbers of editorial pages devoted to various classifications yield useful impressions of relative editorial emphasis. Perhaps of greatest personal interest to prospective authors, however, is a subsection on the editors themselves. Names, titles, education, and experience of staffers permit indentification of a publication as a living thing with which you can communicate.

*Ask the Editor.* Editorial staff members, of course, are direct sources of information about their audience. They spend all their writing time, and more time besides, striving to know their readers better. They use the methods already described, and they meet many readers face to face in regular editorial contacts and at conventions and other gatherings.

The editor is likely to furnish you the aids described above, and he may very well have prepared a special booklet written specifically for you, the prospective author. The editor not only will tell you as much about your audience as he conveniently can, but also will backstop you when he edits your article. If you give too much elementary information, he will tone it down; if he feels you're too cryptic for the audience, he'll ask you to elucidate certain points.

## TREAT YOUR AUDIENCE RIGHT

A few generalizations can be made about the readers of technical and business magazines—at least the business magazines that are aimed at finite areas of technology and business. They are experts in your field, and they are critical. They also are pressed for time, so they must not be expected to dig for their facts.

Your writing must catch their attention and then their interest at the very beginning of your story. A usual way of doing this is to promise data that will be professionally beneficial.

Then follow through. For these readers, no journalistically slick introduction will overcome lack of substance. What may go over big in general magazines—and this includes general business magazines—may fall flat with your expert audience. Probably you've experienced the letdown that occurs when a general magazine you've respected pulls a boner in your field. You realize that the mistake is plausibly presented and you also wonder how much hooey the publication is reporting from other fields of expertise.

Fortunately—once again—the magazine you're writing for has editors who are experts, too. They won't let you make embarrassing breaks, non sequiturs, or unfulfilled promises. Or at least they won't if they've adapted to the traditions of technical and trade journalism and they've been able to stay in the business.

For a close look at the editors, and the journals they shape (or are shaped by), read the next two chapters.

# 5

# THE EDITOR

Visualizing the audience—an exercise that you should practice throughout each writing project—is a basic day-to-day function of each member of a magazine's editorial staff, whether he is editing contributed articles or writing up material that he himself has gathered. Be he an Upper-Case (Editor, Chief Editor, Editor-in-Chief) or lower-case editor (managing editor, associate editor, editorial assistant, and a variety of specialist editors), he is an "editor," and he does his best to understand the readers, because service to readers is the publication's license for existence. Knowledge of readers' needs and wants is a prerequisite to reader satisfaction, which in turn sustains the readers' desire to receive the magazine.

## HE KNOWS THE SUBJECT

Knowledge of the subject matter of the field served is another essential attribute of the editor, as is proficiency in writing and editing.

### Journalist or Technologist?

Which is more important for the would-be editor—knowledge of the field served or journalistic skill—has long been the subject of debate in technical and business magazine circles. The side chosen by an individual has tended to depend on his own background, the technically trained person believing that journalistic skill can be grafted onto a receptive technical person and the journalist contending that he can "learn" any subject area.

A look at the backgrounds of individual editors on a variety of publications permits some generalizations that are pertinent to the debate. First, a mixture of journalists and people with education and/or experience in the field served is found on practically every editorial staff. Second, the more "technical" the field served is, the higher the proportion of nonjournalists. Third, on the more technical publications, the higher positions tend to be held exclusively by people from the field served.

Some deviations from these concepts appear. Journalism predominates in *Oceanology International,* a new magazine for an emerging and largely technical field. Another technical publisher states that his magazine has no staff of technical people, just a diversity of contributors. In justice to the foregoing generalizations, it should be stated that this publication is one of an adjunct breed, a class having tabloid newspaper format that leans heavily on new-product and manufacurers' literature releases, thus explaining at least in part the "diversity of contributors."

Regardless of its source, however, staff competence in the field served is essential, and it is far from new in publishers' tenets. Long ago James H. McGraw insisted that anyone on a paper should know that paper's industry or business backward and forward. In the Gay Nineties he chose the editors of his *Street Railway Journal* for their familiarity not merely with traction but with every phase of electrical development over which the country's greatest industries were becoming concerned.[1] In the long run, McGraw's editors and those of the successor McGraw-Hill, Inc., covered many more fields than electricity, so his principles must have been valid to have generated such a flourishing enterprise.

## HE GETS AROUND

Mr. McGraw also believed that editors should wear out the soles of their shoes, not the seats of their pants. They should be where the action is; go into the mines and mills; live with the jobs; see for themselves; keep in touch with their subjects.

To the chair-bound worker in other lines, such a management tenet might seem heaven-sent, but it's a tough one for an editor to fulfill.

### It's Hard for Most Editors

Writing and editing, planning, corresponding, and doing a hundred little things would keep an editor at his desk all the time if he didn't insist on getting into the field. An average of 1 week of travel a month is as much as most editors can manage, although unusual combinations of opportunity and inclination do permit higher performance.

[1] R. Burlingame, *Endless Frontiers* (New York: McGraw-Hill, 1959), p. 111.

One exceptional construction editor finally managed to push his travel-time factor up close to 75% (almost 3 out of every 4 weeks on the road). He loved his industry, he thrived on travel, he was an efficient worker, and he had good assistants. He used to say with mock ruefulness, "The only trouble with this business is that sooner or later you have to sit down and write."

### It's Still Harder for Many

The need to get around is met less strenuously on some publications that depend heavily on contributed material. Association and technical society magazines tend to fall into this category, and it's a difficult one for a genuine editor to live with. Not only does he miss the freedom to nose around the field, to spot trends, and to crystallize them for all to see, but he also chafes under the yoke of a multiplicity of bosses (members who presume publisher authority).

### One Editor's Story

Hal Hunt, an experienced engineer-jornalist who joined the staff of ASCE's *Civil Engineering* as its chief editor late in his career, speaks on the point of multisupervision:

"Editors of association publications learn (or at least they had better learn) quickly what type of material the executive officers and committees will require that they publish if an author appeals an editor's rejection." [2]

Forestalling appeals became an intriguing game to Mr. Hunt. His favorite gambit derived from the fact that he had a professional engineer's license. When a contributor showed signs of annoyance at an impending editorial decision, Mr. Hunt would point out that he and the author (also a professional engineer) were practicing engineering in different specialties, Mr. Hunt's specialty being the editing of engineering information. He then would suggest tactfully the wisdom of both himself and the author sticking with their own specialties and not attempting to practice each other's.

## HE KNOWS PEOPLE

In fairness to Mr. Hunt's background, his editorship also benefited from his full qualification as a heavy-construction engineer with years of service

[2] H. W. Hunt, "Editing Material for Technical Publications," Paper presented to the Society of National Association Publications (Washington, D.C., 1965).

in both contractors' and owners' organizations. He also had been the construction editor of a "commercial" magazine. With this background he had a wealth of personal contacts to help him maintain his "feel" for the field. A phone call, brief encounters in the halls at society conventions, occasional personal correspondence: these were techniques sufficient to the task.

For an editor, the build-up of cordial and valuable contacts in his field is practically automatic. Whether he goes after a story or it seeks him, his source finds publication an agreeable experience. The editor's part, too, is appreciated, and if the editor's personality is at all warm, friendship logically and usually ensues.

Successful editors for the most part are both warm and outgoing, these characteristics being patently useful in developing confidence, eliciting information, and perpetuating contacts.

### He's a "Loner," Too

These characteristics don't tell the whole story, of course. Many an editor ruefully recalls the promising new staffer who came back from the field, told his fellow workers the fine stories he'd gathered (in complete and articulate detail), but couldn't bring himself to the point of writing the first line. The editor now wonders what ever happened to that happy fellow, and reflects on one other essential characteristic: willingness and ability to retreat from distractions and to concentrate on writing.

Thus an editorial worker needs to supplement alert sociability with lonely dedication to the betterment of his profession.

## PRESTIGE IS HIS

Examples are commonplace of proved, even distinguished, accomplishments by editors in the fields they serve. Before he joined the staff of *Engineering and Mining Journal* in 1874, Richard Rothwell was a recognized firedamp expert and the acclaimed hero of some noted mine rescues. Nathaniel Keith, a founder of the American Institute of Electrical Engineers (now IEEE), joined the *Electrical World* staff in the mid-1880s. Fred Colvin of *American Machinist* was loaned to the Springfield Armory during the industrial build-up for World War I to write detailed procedures for rifle manufacture by United States industries. Arthur M. Wellington's text, *The Economic Theory of the Location of Railways,* was the standard work in its field when he became a magazine editor in 1887. His *Engineer-*

*ing News* piledriving formula still commands respect among soil mechanics and foundation engineers.

### He is a Leader

The *Engineering News* formula is an early example of editorial leadership—the kind for which alert editors constantly strive. Publications do not have a monopoly on innovation—far from it—but when they do come up with something good they are ideally situated to obtain a prompt and adequate hearing. Commercial interests will exploit the ideas if they're money makers.

William A. Stocklin, Editor of *Electronics World*, writes:

> "One of our success stories is in connection with Xerox. This Corporation as we know it today is a direct result of an article that appeared in our magazine quite a number of years ago. We ran a story on electrostatic printing and the then-engineer at Xerox carried from that point and obtained exclusive rights to the techniques." [3]

Warren C. Platt, founder of publications serving the oil industry, claimed fosterhood of the drive-in service station idea.

## HE IS ENERGETIC

Closely akin to editorial leadership is editorial enterprise. The preponderance of monthly magazines in the industrial press suggests a leisurely journalistic pace, with ample time for article production and no need for news. Quite to the contrary, most editorial offices in which I have worked or visited have an atmosphere of urgency that defies correlation with the number of days to the next deadline. The spirit of service to field is the soul and motivation of every one of these magazines, and timeliness comes second only to authenticity in the editors' set of goals. Sometimes in these offices there is excitement matching that of Hecht and MacArthur's classic play, *The Front Page*.

### Another Editor's Story

Such a time for *Engineering News* was the week beginning August 29, 1907, the day the great cantilever bridge that was being built across the St. Lawrence near Quebec collapsed, carrying 74 workers to their death.

[3] W. A. Stocklin, letter to R. H. Dodds (1968).

Full-scale failures present the utmost in challenge to technical journalists, because they are almost certain to embody valuable lessons, while at the same time there is extreme likelihood that one and possibly many entities will want to obscure the lessons. The suppressors fear exposure and the economic consequences of their ignorance or neglect. Great editors conceive it as their highest duty to expand knowledge, so they dig deep—and fast—for the truth before it fades from view.

Young Fred Schmitt (later Chief Editor of *Engineering News-Record*) was the star reporter of the Quebec disaster.

"Editor C. W. Baker was vacationing in Vermont at the time. Immediately he telegraphed the New York office to have associate editor F. E. Schmitt join him at Quebec, since Schmitt had been following the work on the bridge closely. Meeting in Quebec the next morning, the two editors journeyed by trolley and on foot to the bridge, where they found Henry Holgate (head of the board set up by the Canadian government to report on the failure) examining the remains of the lower chord member that had been under observation just before the collapse.

"The following noon Schmitt left for New York with his notes and sketches made on the ground, some photos from the engineer's file (taken before the accident) and some exposures of his own showing details of the collapse. At New York, while these photos were being developed and the sketches copied in a form for linecuts, Schmitt interviewed Theodore Cooper, consultant on the bridge, and prepared his report, spending the night before the next issue went into the mails in the print shop passing copy to the compositors as it was written.

"The result was a detailed report of the accident and a discussion of its probable cause—which subsequently was fully supported in the findings of the official report on the failure." [4]

## Another Failure

This venture, which has been emulated many times by many editors on many magazines, had a rather high-handed forerunner in the 1890s. *Engineering News* Editor Wellington sent a bridge engineer friend to the scene of a bridge failure in Massachusetts. The engineer's analysis pinned the collapse on one structural member, which he managed to knock loose from the wreckage and transport back to New York—as a souvenir, as far as the bystanders knew. Two days later an accurate drawing of the member and an appraisal of the failure appeared in print. Bewildered Massachusetts

[4] "History, Week by Week, Engineering News-Record's Role in Building a Greater America," *Engineering News-Record* (September 1, 1949), A29.

officials talked of legal proceedings against the magazine; however, the officials were placated by return of the member.

## HE IS AN EDUCATOR

Editorial enterprise takes less spectacular forms, too. A good example is the state-of-the-art monograph, tutorial in nature and usually presented as a special section in one issue or as a series of articles. Normally team efforts spearheaded by the staff member who comes nearest to being an expert in the subject, these monographs treat basics (such as lubrication, water treatment, or combustion for the steam-electric power field) as well as up-to-the-minute developments and research (solid-state electronics and servomechanisms for the aerospace field.)

By bringing state-of-the-art together in comprehensive, cohesive units of, say, 16, 24, or 32 pages, the editors may be doing more than compiling fare that is available piecemeal in back issues, and for stable art (does such exist?) they may merely mirror existing material, but in fast-moving technologies the monographs supplement obsolescent texts.

This text-supplementing concept is not new. In commenting on magazine publishing in the late nineteenth century, Roger Burlingame wrote:

"There came, therefore, in this time of transition from trial-and-error practice into what was later called the *scientific method,* vast scope for the current printed word. Books could no longer do the job. Things were moving too fast." [5]

### Extracurricular Activities

The technical editor's essential function of education quite frequently finds expression beyond the pages of his magazine. Many are lecturers in evening schools, some go on to professional posts, and most take part in the activities of their professional societies. In fact, one of the major engineering organizations, the American Society of Mechanical Engineers, was founded in the offices of *American Machinist.* In 1880 a Cornell professor, John Edward Sweet, induced the *Machinist*'s editor, Jackson Bailey, to host the ASME's organization meeting.

Another *Machinist* editor, Kenneth Condit, some 60 years later went on to become dean of engineering at Princeton University. As for ASME, at least one editor, Louis N. Rowley, of *Power,* has served as national president (1967). Waldo G. Bowman, Editor of *Engineering News-Record,* was

[5] R. Burlingame, *Endless Frontiers,* 134.

president of the American Society of Civil Engineers in 1964. William S. Foster, Editor of *American City,* and Edward J. Cleary, a former *Engineering News-Record* editor who became executive director of the Ohio River Valley Water Sanitation Commission during its major growth years, were presidents of the American Public Works Association. William H. Wisely, Executive Secretary of ASCE, was Editor of *Sewage Works Journal,* the organ of what is now the Water Pollution Control Federation.

### Awards, Too

High office is not the only kind of recognition accorded to editors for their leadership. Awards, both formal and informal, come their way. John P. Kushnerick, Editor of *Motor Age,* was honored in 1967 for outstanding contributions to the automotive industry by the Automotive Parts Rebuilders Association's Roy Weldon Award. Frank G. Steinbach, Editor of *Foundry* from 1936 to 1960 and then publisher and editorial director until his retirement in 1967, received numerous medals and awards from associations in his field and was called "Mr. Foundry Industry" by many. Within business journalism he received the first American Business Press G. D. Crain Jr. Foundation Award, which recognizes sustained achievements. Individual achievements (best editorial, best special issue, best contributed article, etc.) are recognized each year by a number of Jesse H. Neal Awards, also sponsored by the American Business Press.

Editors also receive recognition in educational circles. R. Bruce Holmgren, Editor of *Package Engineering,* has won the Packaging Man of the Year Award of Michigan State University for his efforts on behalf of formal education in the packaging field.

## EDITORS ARE IN DEMAND BY INDUSTRY

With all the recognition editors get for carrying out their leadership roles, it is not surprising that many move on to other areas of distinguished service. A few examples have been cited already. W. W. DeBerard, for many years the Western Editor (Chicago) of *Engineering News-Record,* compared his position to that of a goldfish in a bowl, saying that hardly a season of the year passed without some attractive job opportunity being presented to him. Proof came when he retired from the staff in 1940 and promptly became City Engineer of Chicago. His boss, Fred Schmitt, retired the same year and became the economic consultant to the Chief Engineer of the U.S. Bureau of Reclamation.

Donald G. Fink was Editor of *Electronics* before he headed research activities for Philco Corporation and subsequently became Managing Director of the Institute of Electrical and Electronics Engineers. S. Paul Johnston, Director of the Institute of Aeronautical Sciences for many years, was Editor of *Aviation* in the 1930s, and he wrote highly informative articles for the general press in those prewar years. Fisher Black of *Electrical World* went to the Tampa Electric Company, later becoming its president.

## INDUSTRY FURNISHES EDITORIAL TALENT

The foregoing are examples of industry drawing top manpower from magazines. The reverse of this process occurs once in a while, as in the case of *Water and Wastes Engineering* engaging a widely known consulting engineer, Dr. George Symons, as its editor, but a more common device for bringing leaders into the editorial fold is to obtain their assistance on a consulting basis. Some magazines have one or two consulting editors. A few others form editorial advisory boards made up of distinguished industry people.

A typical consulting editor's function is to conduct a regular column in his special field; another is to advise on editorial matters as the need arises. Individual members of advisory boards may function in the latter manner, too, while others may merely accord the prestige of their names.

The most tried-and-true means of getting top brain work from the field, however, is to encourage article contributions. Consistent with the observations of Chapter 2, if a contributor isn't "distinguished" when he starts contributing, he's likely to develop distinction by the time several of his articles have been published.

### Contributor Becomes Publisher

A man who probably was one of the really obscure contributors of all time became a great name in business publishing. His first piece, received by *American Machinist* in 1885, was signed, "J. A. Hill," with no other identification. It was about oiling a locomotive, and it was followed by a succession of cogent letters and articles, all on railroad engineering, that coincidentally revealed the author to be an engineman on the Denver & Rio Grande Railroad. By 1888, the articles, reputedly written from the locomotive cab, were running in almost every issue. In this year an opening on the *Machinist* staff was offered to Mr. Hill, who accepted and came to New York. Ten years later John A. Hill was a publisher in his own right. In 1917, a year after his death, his company merged with McGraw's to form

McGraw-Hill Publishing Company. (Their book departments had merged in 1909.)

Another example of the long leap from contributor to publisher is Milt Kiver. A consultant in the television service field, he was a regular contributor to electronics magazines. Then he became editor of *Electrical Design News,* and finally he started *Electronic Packaging and Production.*[6]

### Stringer Becomes Editor

A special case of the contributor-to-journalist move is that of the regular correspondent, or "stringer," who joins the editorial staff on a full-time basis. Such a person was J. Roland Carr, Western Editor of *Engineering News-Record* until his retirement in 1968. Before 1947 Mr. Carr was a Wyoming Highway Department bridge engineer who supplemented his income by reporting news and technical items from his region. His ability to reap first-rate material from his thinly populated state's rocks and arid soil impressed the editors so much that they persuaded him to join the staff. Of course, a regular correspondent already is a quasimember of an editorial staff, so the move to full-time status is not a long one.

## A FEW CONTRIBUTORS ARE PROLIFIC

Some contributors of articles about their own area of expertise achieve a kind of editorial staff status, merely by the regularity of appearance of their by-lines. Such a man is Igor J. Karassik, of the Worthington Corporation, who in 1966 was feted at a luncheon hosted by *Power* on the occasion of his 1000th contribution to technical literature. These contributions were published by 105 magazines in the United States and abroad.

James J. O'Connor, executive editor of *Power* at the time, wrote in part:

"Obviously Karassik likes to write. And his easy style sits well with his engineer readers. He tackles writing with zest, shows marked ability to identify areas that can profit from editorial coverage at the time.

"Back in the early 1950s, Igor dropped into our offices to discuss the practicability of someday being able to vary pump speed by changing motor frequency through use of solid-state frequency changers. This is typical of his engineering vision, which he then develops in his writings for a far-flung engineering audience.

"What are some tangible benefits from such an extensive writing program? A worldwide reputation as an authority on pumping is one. Also,

[6] W. A. Stocklin, letter to R. H. Dodds (1968).

Igor tells me, college expenses for his four children matched writing income."[7]

Another prolific writer was Dr. Gustav Egloff of Universal Oil Products Company. A 1947 *Science Illustrated* article attributed nine books, 500 articles, and 300 patents to Dr. Egloff, and called him a "walking encyclopedia of petroleum science and technology." The editor of a publication in the petroleum field made the following curious comment on Dr. Egloff: "Although he was not our most prolific writer, he did have the reputation of writing more than he read."

## A Versatile Engineer-Writer

Among practicing civil engineers the late D. B. Steinman had no writing peers that I know of. Others were prolific in one or two specialties: Homer Hadley in structures, Karl Terzaghi in soil mechanics, Willem Rudolfs in sanitary engineering, to name only a few. But Dr. Steinman's facile pen flowed on a variety of subjects throughout his productive career. A bridge engineer first, he also was an early and persistent pusher of professionalism for engineers, and he wrote extensively in both these areas. His book, *The Builders of the Bridge,* about the Roeblings and the Brooklyn Bridge, is a first-rate historical writing job that imparts the author's enthusiasm and love for his subject.[8] He wrote about the bridges that his firm designed, and he wrote still more about bridges that ought to be built—some of which he eventually engineered and some of which ultimately were designed by others.

Dr. Steinman was a long-time member of the New York State Board of Examiners for Professional Engineers and Land Surveyors, and he did much through his writing to extend registration to other states and to develop uniform registration standards among the states. Allied with this movement was the organization of the National Society of Professional Engineers. In the 1930s Dr. Steinman visited influential engineers in state after state and wrote tirelessly for any publication that he believed could get the NSPE message to interested persons. His "The Engineer," a lyric impressively reprinted for framing, hung on the walls of many engineers' offices throughout the United States.[9]

Dr. Steinman did not wait for magazines to ask for his writings. He just sent them in to whichever publications he felt should use them. Donald D. King, who was editor of *Civil Engineering* during part of his career on the

[7] J. J. O'Connor, *Power* (April 1966), 2.

[8] D. B. Steinman, *The Builders of the Bridge* (New York: Harcourt, Brace, 1945).

[9] D. B. Steinman, "The Engineer," unpaginated reprint from *Civil Engineering* (December 1942).

ASCE staff, once said, "Dr. Steinman would fill up every issue if we were willing."

## THEY TELL THE WORLD

The influence of versatile writers like Dr. Steinman and many business-press editors usually extends beyond their own fields, and at the least they perform the valuable service of interpreting their fields to outsiders.

One more example is pertinent. M. D. Morris, an alumnus of Cornell University and a professional engineer, has conducted a profitable sideline as a free-lance writer on engineering subjects for years. More than 300 of his articles, many of them for general magazines, and most about the work of others, have been published. He also is a consulting editor for John Wiley & Sons, and teaches courses in communications at several colleges.

## AN EDITORIAL CAREER?

Mr. Morris probably prefers his unique way of life, but his is a type of talent and interest that make good editorial staff material for magazines. Which brings up a question that anyone who likes to write should ask himself: "Would an editorial career be a good one for me?"

For those who look beyond the hard work to peripheral matters like pay and benefits, the following observations are pertinent. Publications try to pay their editorial staff members as well as or better than they would be paid if they held positions of comparable maturity in the field served. Magazines in the larger publishing houses have benefits (vacations, insurance, and retirement plans) on a par with those of the largest industrial companies. Smaller houses tend to follow suit within their financial capabilities.

And what of the work itself? Would it be to an individual's liking? Much of the answer can be found in his reaction to this chapter and to the next one, to the extent that they tell what an editor does and what a publication is.

# 6

# THE PUBLICATION

Technical and business magazines may acquire recognizable personality traits from individual editors, both present and past, but the successful publication is shaped primarily by external conditions. The most influential condition is the informational need of the field served. In second place is publishing practice, which includes a range of factors from ethics to format, and which has developed a breed of publications that for the most part behave and look alike.

## IDEALS

James H. McGraw, surely one of the most influential developers of the breed, wrote on the ideals of industrial journalism:

"First among these ideals is independence—the determination of a journal to be its own master, to have no other guides for its opinions and policies but truth and the sound interests of the field it serves. The right-minded publisher holds that he has a covenant with his subscribers—a covenant to be honest with all and to do harm to no one who is pursuing an honest course. From that covenant he will not depart.

"An industrial journal so controlled at once commands respect. If it is to exert influence as well, it must also command ability of a higher order. Managerial skill, necessary as it is, will not alone suffice. There must be that rare editorial power, compounded of knowledge, alertness and foresight, that can keep its finger on the pulse of an art and an industry, analyze achievements and comprehend tendencies, that . . . mistakes neither charlatanism for genius nor genius for charlatanism. Not even the editor of an industrial paper, of course, can be infallible, and when he is in doubt he does right honestly to say so; but it is his part to bring to his task of interpretation and guidance all that industry, application and constant contact with his field can supply." [1]

## Editorial Independence Comes First

Independence, in Mr. McGraw's context, clearly includes independence from improper influence by advertisers. At the risk of appearing to protest too much, I assert that every right-thinking and continuously successful publisher hews strictly to this line.

Perhaps surprisingly, it is not difficult to maintain this particular form of independence, because industrial journals possess broad community of interest among advertisers, editors, and readers. In fact, advertisements complement editorial material to fill out the over-all informational package. However, advertisers occasionally do make unreasonable requests, either through space salesmen or directly to editors. But such is the alertness of editors and such are the principles of publishers that the requests are either turned aside or transformed into beneficial results. A commonplace in industrial journalism is the editor's account of an offended advertiser who withdrew from the magazine, only to return to the fold with still greater confidence in the magazine's usefulness to him.

## Advertising Must Pay the Bills

There is no intention here to belittle advertising. Not only does it round out the informational package; still more importantly, it pays the bills—for office rent, salaries, editorial travel, and first-rate printing. That such an obvious fact was not always known is a little hard to understand, but a principal forebear of one of our most important magazines nearly foundered on this very ignorance. Today's magazine is *Engineering News-Record,* and the forebear was *Engineer and Surveyor,* later *Engineering News.*

George H. Frost, a surveyor, started the paper in 1874, depending on subscriptions from his chosen audience, surveyors, to pay the publication costs.

". . . Frost overlooked [wrote Henry H. Norris] the fundamental publishing axiom that to be self-supporting a publication must serve readers who control reasonably large expenditures for supplies in its field. Surveyors were not in this class." [2]

[1] J. H. McGraw, "Ideals of Industrial Journalism," *Electrical World* (September 20, 1924), 559.

[2] H. H. Norris, The Story of the *Engineering News* (unpublished manuscript in McGraw-Hill archives), p. 3.

Roger Burlingame comments on this quote:

"Mr. Norris, of course, was writing from hindsight. In 1874 what he states as an axiom was scarcely self-evident to many publishers. Benjamin Franklin had realized it more than a century before, but in many fields Franklin was thinking at least a hundred years ahead of his time." [3]

## HISTORY

Mr. Franklin's contribution to journalism was the newspaper for all people. He founded the *Pennsylvania Gazette* in 1728. It is appropriate, therefore, that the first publication of general engineering flavor was the *Journal of the Franklin Institute,* first published under this name in 1826 after a year as the *American Mechanics Magazine of Philadelphia.* Frank Luther Mott considered the *Journal* one of the first and greatest know-how magazines.[4] Before that, in 1797, the quarterly *Medical Repository* was founded under the editorship of Samuel Latham Mitchill, a chemistry professor at Columbia University, and by Dr. Edward Miller.[5] Another early specialized scientific publication was the *American Mineralogical Journal* (1810, New York). Archibald Bruce, a mineralogist, was its editor.[6]

The first number of the *American Journal of Science* (New Haven) appeared in July 1818. Benjamin Silliman, an outstanding scientist, was the editor.[7]

### Railroad Magazines Were Early Specialists

Beginning in the 1830s and continuing throughout the remainder of the nineteenth century, many railroad magazines were published. Demonstrating the point that specialized publications arise, flourish, and even change their emphases in response to specialized activity, the early railroad-magazine movement coincided with the great era of railroad expansion. Subject matter emphasis was on construction as well as equipment and operation. Later, when building decelerated and reached a practical standstill in the early years of this century, the railroad magazines gave less and less attention to the civil engineering phases of their field. One important railroad magazine genealogy is illustrated in Figure 1.

[3] R. Burlingame, *Endless Frontiers* (New York: McGraw-Hill, 1959), p. 50.

[4] F. L. Mott, *A History of American Magazines* (Cambridge: Harvard University Press, 1938), Vol. 1, p. 445.

[5] *Ibid.,* p. 149.

[6] *Ibid.,* p. 266.

[7] *Ibid.,* p. 303.

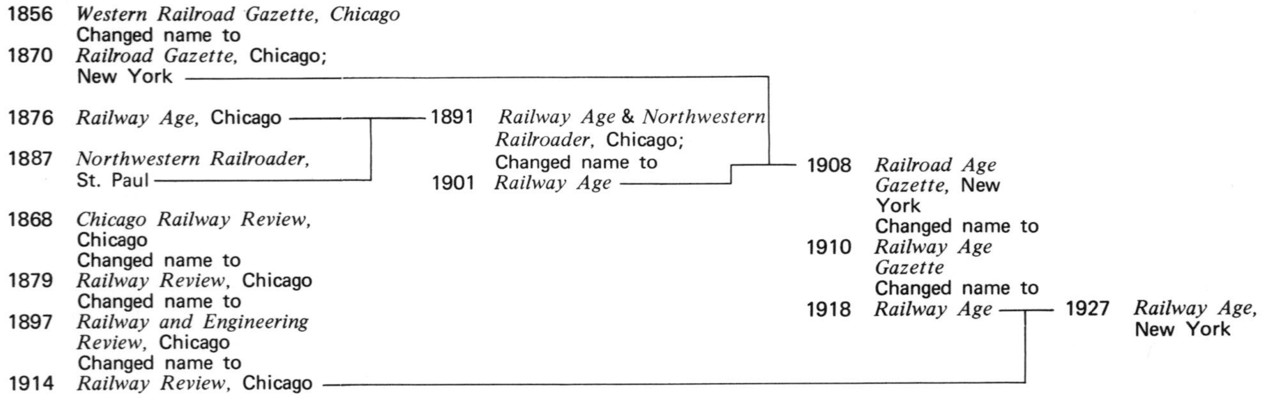

1856 *Western Railroad Gazette, Chicago*
Changed name to
1870 *Railroad Gazette,* Chicago;
New York

1876 *Railway Age,* Chicago

1887 *Northwestern Railroader,*
St. Paul

1868 *Chicago Railway Review,*
Chicago
Changed name to
1879 *Railway Review,* Chicago
Changed name to
1897 *Railway and Engineering*
*Review,* Chicago
Changed name to
1914 *Railway Review,* Chicago

1891 *Railway Age* & *Northwestern*
*Railroader,* Chicago;
Changed name to
1901 *Railway Age*

1908 *Railroad Age*
*Gazette,* New
York
Changed name to
1910 *Railway Age*
*Gazette*
Changed name to
1918 *Railway Age*

1927 *Railway Age,*
New York

**Figure 1. Genealogy of Railway Age.**

*Eclectic Engineering Magazine* (1869, New York) is an example of subject-matter evolution. David Van Nostrand, a civil engineer and founder of the technical book company that bears his name, published *Eclectic* until his death in 1886. At that time the magazine, whose name in 1879 had been changed to *Van Nostrand's Engineering Magazine,* passed to *American Railroad Journal.* Although the editorial content was substantially civil engineering at first, mechanical engineering received an increasing share of the space as the years went on.

From a journalistic standpoint, Van Nostrand's magazine is of interest because it was an eclectic of a newer ilk (in older days many editors leaned heavily on material borrowed from other publications—frequently without crediting the source, because literary piracy was not a crime.)[8] Its policy was stated in the first issue; in effect:

> This magazine is eclectic in the proper sense. It does not consist of slices and reprints. Rather it consists of abstracts of important engineering writings from the press of the World. Further, condensation must be more by pen than by scissors. Our purpose is to save the engineer his valuable time.[9]

By the time the word "eclectic" was dropped from Van Nostrand's title, the magazine contained original articles, but digests of articles, especially foreign articles, continued to appear through 1886.

## Professional Societies Fostered Journals

*Eclectic*'s early years practically coincided with another development that reflected advancing technology. In 1867 the American Society of Civil Engineers (founded in 1852) began publishing its *Transactions,* followed in 1873 by the companion *Proceedings.* As successive engineering societies were founded, they developed similar publications. Together with older titles mentioned above, these periodicals are categorized as technical and scientific journals, a viable form of publication that grew apart from the technical and business magazines. The journals, which present entirely contributed material to subscribers who usually are members of the sponsoring organization, and which do not necessarily carry advertising, are treated in *Writing for Technical and Professional Journals,* by J. H. Mitchell.[10]

[8] *Ibid.,* p. 39.

[9] *Eclectic Engineering Magazine* (January 1869), 1.

[10] J. H. Mitchell, *Writing for Technical and Professional Journals* (New York: Wiley, 1968).

## Advertising Policies Emerged

The place of advertising in the infant industrial press was but dimly understood in the post-Civil War years. It took the genius of A. M. Wellington, who joined Frost's struggling *Engineering News* 13 years after its founding, to develop a real community of interest between this magazine's readers and advertisers. On the other hand, Henry C. Meyer, the founder of *Engineering Record* (née *The Plumber and Sanitary Engineer,* 1877), was a dealer in plumbing supplies who grasped immediately the value of a technology-hungry audience as a coincidental target for advertising. He also understood the promotional value of leadership, sponsoring in the second year of the magazine's existence a contest for the best design of a tenement house to be built on a standard city lot of 25 × 100 feet.[11]

One bit of philosophy on advertising that has been called "unique in publishing history" was that of *American Machinist*'s founding editors, Jackson Bailey and Horace B. Miller. In the leading editorial of their first issue (1877) they stated, "We are not specially interested in the sale of machinery or other merchandise." They set a rigid limit of 16 pages per issue, ran all the editorial material they considered useful, and filled whatever was left with advertisements. From the beginning they had an advertisers' waiting list. (They also managed to create a paid subscription list of 5000 before the first issue went to press.)[12]

In the last decade of the nineteenth century and the early years of the twentieth, advertising philosophies and publishing practices as a whole crystallized to a strong resemblance of those of the technical and business magazine press we know today. Rightly, editorial excellence came first. It was followed by aggressive promotion of advertising space sales and, ultimately, by concepts of circulation purity (with strong overtones of "truth in advertising").

## Multipublication Houses Grew

Groups of magazines under individual or corporate ownership also developed. Some groups, like Warren Platt's oil papers (Cleveland), served one industry. Others (McGraw-Hill, Chilton, Cahners, and Reuben H. Donnelley, for example) served two or more industries, the objectives or advantages of grouping being to gain a variety of economies and to develop broader profit bases.

[11] *Engineering News-Record* (September 1, 1949), A25.
[12] R. Burlingame, *Endless Frontiers,* p. 89.

## Big-House Economies

Economies available through grouping include centralized office services, accounting, advertising sales, subscription records, art department, and magazine production. The last-named is a centralized receiving and shipping function at the interface between each magazine's staff and the printer. This function does not perform any editorial work. In fact, each big-house editorial office resembles its outside competitor's office more nearly than it resembles other editorial offices in the same company. A magazine's autonomy is most clearly visible in its editorial office, and big-house advantages are least perceptible there.

Advantages do exist, however. Perhaps the most important advantage is access to a number of experienced people in management and on the other papers who can advise on innovative ideas, warn of pitfalls, and assist staff development.

Notice that it is *access to* advice and assistance that is an advantage; the initiative should be and usually is taken by the individual publication. Gratuitous suggestions from outside a publication's own editorial offices can be offensive to its jealously guarded autonomy. Of course, in multipublication companies conflicts can arise out of overlaps between fields served. Magazine A, serving the construction industry (and therefore being in favor of anything that fosters construction activity), editorializes on behalf of a proposed public power project; magazine B, in the same company, serves investor-owned utilities, who oppose public power. The editor of magazine B might want magazine A to soft-pedal the editorial, but he would be out of order if he spoke up. In matters of editorial conflict between fields served, the fact that two publications are in the same house is purely coincidental.

## Big-House Profit Bases

Another big-house advantage is the availability of financial backing for editorial projects intended to improve a magazine's long-run position (and profitability). Similarly, a multimagazine publisher can start a publication in an emerging field, offsetting initial losses by profits from the established papers.

Ultimate profitability is a sine qua non for a new magazine's success, and no substitute has arisen for the principle Mr. Norris stated in his comment on the early misdirection of the *Engineer and Surveyor*. Another example of a publication that did not serve readers "who control reasonably large expenditures for supplies" was *Firemen's Journal* (New York,

1877). Here again opportune redirection saved the day. Fire fighting required large amounts of water wherever a fire could occur. If money was available to pay for delivery of the water, adequate protection could be assured. Thus water supply and distribution was a major expenditure area in fire protection, and in 1886 the *Firemen's Journal* changed its name to *Fire and Water.* It also became the first commercial public works magazine, an ancestor of today's *Water and Wastes Engineering,* published in New York by Reuben H. Donnelley.

## Where the Spenders Are

General subject categories that involve readers who control reasonably large expenditures for supplies are engineering, technology, and most applied sciences, together with the functions of business administration, education, and government. Social science, humanities, and life sciences do not support commercial publications to any significant extent, although ample literature is developed through the medium of professional journals.

Individual magazines deliberately overlap categories, a common type of overlap being the juxtaposition of business administration and engineering. A publication covering a field such as metalworking will be concerned with management, marketing, research and development, and production, treating the last-named in all its engineering aspects: design, construction, tooling, automation, operations, and packaging.

## Horizontal and Vertical Publications

The terms "horizontal" and "vertical" frequently are applied to technical and business magazines, the former implying a function (e.g., design, accounting, or advertising) that is common to a number of industries, and the latter treating an industry from bottom to top (the "overlap" example above). Pure examples of either term are difficult to discern, but the concept is useful for defining readership, slanting stories, and selecting advertising media.

## What's in the Name

A magazine's name usually gives a good idea of the field it serves, as inspection of the titles in Appendix C will demonstrate. This list comprises United States publications that use contributed articles and have at least nationwide distribution. For the most part, circulations exceed 10,000, the prevailing range being 25,000–50,000 and some being well over 100,000. Most have some international distribution and a few are primarily for overseas readers.

Most magazines for foreign distribution are published in English, although a few are in Spanish. *Ingeniería Internacional,* started in 1919, was an early example. The overseas magazines carry the advertising of U.S. producers who distribute to foreign markets.

### Specialized Sponsorships

Some publications are owned by membership societies that also publish professional journals. The society magazines are intended to be self-supporting, and indeed many of them bring a handsome advertising revenue that helps defray over-all organization costs.

An unusual type of sponsorship is supplied by the industrial company. *Compressed Air Magazine,* founded as *Compressed Air* in 1896 by an official of Ingersoll-Sargent Drill Company, is an outstanding example. Now published by Compressed Air Magazine Company, it has been under the wing of the Ingersoll organization (now Ingersoll-Rand Company) continuously. Ingersoll-Rand advertising is conspicuous but by no means predominant in *Compressed Air Magazine.*

### The Regionals

Not listed in Appendix C are regional and statewide magazines, many of which welcome contributed material. In construction journalism, for example, about a dozen "regionals" cover the United States. For the most part they maintain nonoverlapping geographical coverage as far as circulation is concerned, and editorial coverage strongly emphasizes activities within the respective areas. An association of regional construction papers presents their readership data to advertisers, and the members offer combination rates to advertisers who take space in all papers.

## THE INFLUENCE OF MAGAZINE ASSOCIATIONS

Voluntary association among magazines is not confined to regionals. In fact, one of the most important reasons for the success and prestige of today's business press is the existence of so-called trade associations in the field. Thanks to the activities of these associations anyone can obtain detailed data on a member magazine's circulation: where it goes and to what types of buying influences. (Standardized circulation statement forms are described in Chapter 4.)

Moreover, every publication prints a standard advertising rate card, so that all advertisers come to the marketplace on an equal footing.

This fair-and-square situation did not always exist. In the late 1800s, when the money-making potential of technical and business magazines began to be realized, opportunistic space salesmen charged whatever the traffic would bear—usually at some haggled-to amount below an adequate but unpublished asking price. Part of the "solicitor's" pitch was a glowing description of his publication's readership. Circulation was typically claimed to be much higher than it actually was, occasionally to the subsequent embarrassment of this or that publisher who happened to have progressive scruples.

A few magazines attempted to counteract their advertising representatives' exaggerative tendencies by printing their circulation figures in each issue. This was the genesis of today's independently audited circulation statements.

### The Origin of ABP

Widespread adoption of the leaders' principles was not spontaneous, however, and it took an unrelated event to bring the publishers together. In 1905 a pending federal bill threatened to double the second-class mail rate. To form a bloc in opposition to the rate rise, publishers of engineering, manufacturing, and merchandising magazines met at Niagara Falls. The postal rate increase did not pass, presumably as a result of this and other concerted efforts. The Niagara Falls meeting also inspired the founding of the Federation of Trade Press Associations. (There were several local associations in existence). A few years later, reflecting both a desire for cohesive permanence and a growing distaste among the publishers for the term "trade," the organization became Associated Business Publications, Inc. (ABP, now American Business Press).

## MODERN PRACTICE

Over the years, ABP and other interested organizations of industrial advertisers and circulation-audit people have fostered and brought into widespread acceptance the practices pioneered by the early leaders. These practices may be grouped into three areas: circulation, advertising, and editorial.

### Circulation: Paid versus Free

Of the three, circulation is taking the longest to mature. The problem is akin to the writer's and editor's need to know the audience, but it is sired by the advertiser's insistence on knowing how many of his customers and prospects sit in that audience.

An audited circulation statement presents facts. If those facts do not satisfy enough advertisers, the paper either folds or takes steps to make the facts more palatable. Redirection of editorial effort is a time-honored step, as we have seen, but if the editorial effort already is aimed at the people who will buy the advertiser's wares, then the circulation effort must be modified. Subscription salesmen confine their calls to classes of people the advertisers want, and direct-mail sales lists are refined with the aid of appropriate directories and canvasses of known "desirables." Furthermore, each subscription order must identify its maker as being "qualified," that is, within the intended audience.

So far so good, and circulation practices would have been stable long ago if it had not been for the ascendance of "controlled circulation," circulation comprising recipients carefully selected for their buying influence. Controlled circulation was long scorned as "free" or "throwaway" by paid-subscription publishers, and some of the scorn was justified. Advertisers' mailing lists became magazine circulation lists, a practice that at least suggested the possibility that editorial independence was not all it should be on such magazines. Against this abhorrent concept, the paid-subscription magazines urged the self-evident superiority of the paper that a man paid for because he wanted it.

As the battle waxed, ABP put out a circular, "The Plus of Paid Circulation," which stated in part:

"Each [publication] has a contract with its subscribers to deliver a *specific* number of issues of a *specific* editorial character for a *specific* length of time at a *specific* price." [13]

Controlled circulation eventually came up with answers to all objections (except objections to the bare fact of nonpayment), and even the die-hard subscription-only houses found that controlled circulation was more effective in some service areas.

Questions about the recipient's desire to receive a controlled-circulation publication are now usually answered by means of a qualification questionnaire, which requires him to state that he wishes (or does not wish) to receive the magazine. The form also requires qualification data (employer, title, function, and relative buying influence) that are at least comparable in value to the qualification data required with a paid-subscription order. The controlled-circulation magazines have an equivalent of the renewal of paid subscriptions, too—requalification. Requalification questionnaires are sent to current recipients at regular intervals, usually a year and a half or

[13] Folder, undated.

two, and failure to return the card (or a second one) results in removal from the list.

Many publications, particularly newer ones established after controlled-circulation practices acquired dignity, began under the no-pay banner. Others, some of them of great age and unquestioned prestige, have switched from paid to free. A notable example is the same *American Machinist* that had a paid subscription list of 5000 before the first issue came out. The commonest reason for switching appears to be the success of free competitors, although exceptionally high subscription sales and service costs could make the decision attractive.

## How Circulation Can Pay Its Way

The days when subscriptions paid all magazine costs—editorial, production, and distribution—are past, even for the notable *Reader's Digest,* and those days practically never existed in the business press. Nevertheless the concept of subscriptions making or losing money is part of the circulation picture. In this concept, every nickel of subscription money received by the publishing company is credited to the circulation department. (Few business publications have significant single-copy sales). Against this account are charged sales commissions, subscription advertising, premiums, direct-mail promotion, and circulation record-keeping costs—in general, all the costs of delivering to the mail room an up-to-date run of address stencils for each issue. If these costs total less than subscription income, the subscriptions are said to make money. Some do, but many don't. When losses exceed the estimated cost of maintaining a controlled-unpaid list, management is likely to consider going to free circulation.

Up to this point we have used the term "controlled" in its popular (controversial) sense, implying free circulation. The free-circulation people scored a coup when they pulled that word into their camp, because "controlled" really applies with equal validity to both unpaid and paid circulation as it is practiced today—and insisted upon by advertisers. This conclusion is readily drawn from the foregoing discussion.

## Advertising: High Prices, Truth, and Dignity

Advertisers pay high prices for space in technical and business magazines. The common quotient for comparing advertising space is cost per thousand readers. This cost for a page in a large-circulation general magazine is in the order of $5.00 per thousand, whereas a specialized magazine may get 5 times as much.

Some idea of the profitability of a publication can be gained by its ratio of advertising pages to total pages. A traditional ideal from the publishing business standpoint is two-thirds advertising, but many magazines survive quite adequately on one-half, and lower ratios are not uncommon.

Industrial advertisers impose one significant but superficial condition on the technical and business magazine: a standard size for a full-page advertisement, resulting in an 8¼ × 11¼ inch trim size for most magazines. The standard size was brought about by the fact that many industrial advertisers place identical advertisements in several magazines; the uniform size permits reuse or duplication of letterpress printing plates, thus eliminating enlargement or reduction costs for printing in different-sized magazines. Increasing use of photo-offset printing, with attendant savings of enlargement and reduction costs, tends to deemphasize the importance of the standard, but its effect is fully established and few deviations are likely to occur in page-ad size.

Two magazine-size variations are used by a few publishers, but they maintain the ad-size standard. One is the square book, approximately 11¼ × 11¼ inches, which runs a column of editorial material alongside each full-page ad; the other is the tabloid newspaper size, which permits editorial material to be placed both beside and above the standard advertisement. Advantages of guaranteed ad placement adjacent to editoral material are asserted for these format variations, the accepted philosophy being that readers take a publication for its editorial content, which leads them to the ads.

It hardly requires mention that ads that look like editorial matter are taboo (unless clearly marked "advertisement"), because this is such a firmly established principle of reputable publishing, but it is only a part of overall magazine business practices seeking to assure that advertising will be on the same high plane with editorial matter and that the former will not govern or even improperly influence the latter. Another practice is insistence on truth and dignity in advertising. Truth in advertising has been well established since the decline of rampant *caveat emptor.* The requirement of dignity in the technical and business press is analogous to certain provisions of codes of ethics in professional societies. For example, the code of the American Society of Civil Engineers states, "[he will not] advertise . . . in self-laudatory language or in any other manner derogatory to the dignity of the profession." Asking a machinery manufacturer not to advertise in a self-laudatory manner may be asking too much, but a publication would be acting within the bounds of business-press propriety if it rejected an ad that tended to discredit a competitor (short of libel, of course). A practical reason for this caution would be the desirability of keeping peace within the industry. (General publications have fewer inhibitions of this nature.)

Monitoring advertising practices is a natural editorial function, because editors are habitually concerned with the over-all excellence of publications. Moreover, with their traditional independence they can view advertising problems objectively and in long-range context.

Timewise, however, keeping an eye on the advertising is but a small part of editorial practice. In a broad sense, most of the other chapters of this book are devoted to the editor's work. The next few paragraphs present just a few comments to round out the picture.

## Editorial: Independence, Sense, and Dollars

These comments are primarily on the financial aspects of the editorial process. Editorial budgets are typically tight in the technical and business press. This does not mean that salaries are low; they must be sufficient to attract first-rate people. But it does mean that the people must work both hard and efficiently; make good use of contributed material; plan their travel carefully for maximum copy and contact yield; make effective use of correspondence and writing-speed aids like Dictaphones; develop a sixth sense for worthwhile reading and a knack for rapid comprehension; avoid nonproductive activities; and read lots of proof outside working hours.

The tight budget is not a whim or happenstance of business-press life. It is built into the economics of low-circulation magazines—low, that is, with respect to the circulations of general magazines. A large-circulation magazine's costs are more sensitive to noneditorial influences: the cost of paper, the cost of mailing, the cost of a new printer's union contract—items that change the per-copy cost by amounts that must be multiplied by hundreds of thousands or millions. For these magazines the addition of a staff member, or an extra trip around the world for a reporter-cameraman team, are but drops in the bucket.

The situation is reversed on a magazine having, say, 40,000 circulation. There the addition of a staff member might take 1 or 2% of the over-all costs and affect profitability by a much larger factor. Production cost changes, on the other hand, would have less significant effects relative to the over-all cost of the low-circulation magazine.

Big magazines have no trouble paying handsome rates for contributed material, either. Perhaps the disparity between their rates and those paid by most business papers explains the fact that the latter prefer the term "honorarium" to "pay" or some other word implying just compensation for the amount of effort involved.

"Honorarium," according to the *American Collegiate Dictionary,* is "1. an honorary reward, as in recognition of professional services on which no price may be set. 2. a fee for services rendered by a professional person."

Actually, the magazines that pay honoraria (technical society magazines don't, as a rule) relate these disbursements to the amount of space occupied by contributed material, and thus, in effect, they set a price on the services. Typical "going rates" today are $25 or $30 a published page, so there is little argument to the effect that the rates compensate adequately for the effort.

Fortunately for both editorial budgets and contributors, by-line prestige is a valuable consideration in the professional arenas, and so people in these fields will continue to do their part on behalf of a vital and sometimes prosperous business press.

Now for the nuts and bolts of contributing.

# 7

# SELLING YOUR ARTICLE

The aggressive idea of "selling" your article may seem incongruous in a journalistic environment, where the word "submission" commonly describes the offer and where an unaccepted submission is "rejected." This negotiating environment suggests domination by a lion of an editor over a lamb of an author.

Lion and lamb coexist congenially, however, when you exercise the selling techniques to be discussed in this chapter. At best you will end up with publication of a fully competent story; at worst you will be a relieved and happy man because you didn't have to go through the agony of writing, only to find out that the editor had no use for your topic.

## CONDITIONAL ACCEPTANCE IS YOUR GOAL

The process—the technique—is essentially a dialogue with the editor. It may begin as a few words exchanged in the corridor during a convention, or it may be handled entirely by correspondence or telephone. The dialogue will be complete when the editor decides whether or not to give you a conditional acceptance—conditional, that is, on the article's getting written and turning out to contain all the information that your exchanges promised. This conditional acceptance is uniquely feasible in the specialized press because usefulness of the information, rather than quality of writing, is the criterion of acceptance. Obtaining conditional acceptance is the essence of the selling technique.

### Select Your Prospect(s)

Understanding and possibly some real work must precede the initial contact. Like any other salesman, you will select your prospects. This function is a form of market analysis, which you may be able to conduct instantly or which may require some conscientious digging. Quick prospect selection is possible when you feel that your topic is appropriate for a magazine you've known for a long time. Almost unconsciously you know who the readers are and the kinds of material the editors like to receive. You also have a good idea of what that magazine has published lately.

On the other hand, if you believe your topic is too specialized or too broad for that magazine, you may feel compelled to get acquainted with some strangers.

Appendix C lists about 300 technical and business magazines, generally national in scope, that publish contributed articles. Checks on the numbers of listed magazines that serve various subject areas yield results that will surprise many professionals in those fields. For example, the areas of food industry, railroads, chemical engineering, and business administration have 7, 6, 11, and 17 publications, respectively. Doubtless magazines having allied or peripheral interests would more than double each of these numbers. Thus it is quite evident that few prospective authors would be thoroughgoing readers of all the publications that conceivably could want their articles.

Before attempting to discover the right publication for your story you should decide what kind or kinds of audience would get the most benefit from the information. This decision not only helps you discharge your professional obligation in the best possible manner but also more nearly assures eventual publication.

Your investigation of publications is likely to turn up more than one prospect. Thus you will be able to establish an order of preference, trying your idea on a second or third magazine if the first one is not interested. Moreover, you may perceive audience differences that suggest various treatments of your subject, each treatment being for a different, interested audience, and each one a separate title in your list of publications.

## Play Fair!

Some words of caution, however: play fair with the editors. Deal with one magazine at a time in the same field; if you believe they're in different fields, tell their editors about your current negotiations with other magazines. There's always a chance of more readership overlap between two publications than you were able to discern, and an editor can become as vengeful as a deceived spouse if he finds out after publication that you placed the same story with a competitor.

### Is Anyone Ahead of You?

Chapter 4, "Your Audience," contains suggestions on learning about publications. Most aspects can be covered with the aid of literature about a magazine (circulation statements, etc.), but to find out whether that magazine has covered your topic you will have to inspect current and back issues (you should do this anyway) or ask the editor. The latter move yields one more bit of necessary information: whether an article similar to yours is already scheduled for a future issue. The move also starts the overt selling process, and, assuming that the editor tells you the topic is open, you should follow immediately with your main pitch.

### Now Make Your Presentation

Describe your topic to the editor; tell him what it is and how it would benefit readers. If you have prepared an outline, so much the better. The more you send, the better the idea the editor can form about eventual acceptability and the more helpful he can be in the way of suggestions. However, your objective at this stage is to get a tentative commitment without waste effort. Don't strain for an outline. The editor will ask for it if he wants it. More likely he will ask you specific questions to point up topic facets he considers important to the readers.

If and when the editor is satisfied by your data, he will give you conditional acceptance; that is, he will tell you to go ahead and write the story.

Of course, he could be so anxious to have your story that he would offer to interview you and write the story for your approval and by-line. A few magazines regularly follow this modus operandi, but for most it is a kind of reserve procedure, available for occasions when early publication is desired and the editors want to be sure the writing gets done on time.

## HOW ACCEPTANCE PRACTICES DIFFER

An opposite extreme in practice is to insist on finished manuscripts only. I know of no publication in the technical and business field that adheres to this practice, but where there is heavy contribution pressure the sheer weight of preliminary correspondence could lead an overworked editorial staff to adopt a prevailing "just show us the article" response to idea suggestions. This response is the rule rather than the exception among general-circulation magazines, which also make wide use of "rejection slips" or form letters when finished articles are not acceptable. In a survey of publications listed in Appendix C, only one magazine, *School Management,* reported the use of a standard letter to accompany returned manuscripts, and this form includes a checklist to indicate the general reasons for rejection. Moreover, the letter states in part, "If you want to revise the article and try again, by all means do. But query us *first* for more detailed suggestions."

Most publications responding to the survey stated that they welcome contributions, and some that went into more detail discussed the practices of preliminary contact.

In an excellent author's guide, *Industrial Research* states:

"To save time and to insure proper editorial approach . . . send the editors an outline, abstract, or rough draft before submitting the finished manuscript."

Editors of the *Harvard Business Review* regularly appraise rough drafts and carefully developed outlines for prospective authors. When the material is not considered suitable for *Harvard Business Review,* the editors may even take the further step of advising the author where he might find acceptance.[1]

*The Electronic Engineer* welcomes outlines but doesn't insist on them, stating in a Ditto sheet answering typical author questions, "An outline will help you write the article, but it's not necessary that you submit one to us. The most we could do with an outline is to express our interest in the subject."

*School Management* comes concisely to the point in a pamphlet presented in question-and-answer form:

**Q.** Will you consider anything but a finished manuscript?

**A.** Yes! In fact we are eager to have you query us before you go to a lot of work. A long, rambling letter that explains the story you have in mind can often help us to help you." [2]

Although it is by no means universal practice, many magazines have some form or other of printed advice to contributors. Another pamphlet, prepared by *Air Conditioning, Heating and Ventilating,* states under the heading, "How to Go About Writing":

"This is a great deal simpler than you might think: in fact it is so simple that it is surprising that so few engineers do write, when you consider the many advantages to them.

[1] S. A. Greyser, "The Harvard Business Review," *Harvard Business School Bulletin* (March–April 1966), p. 7.

[2] *How to Write for School Management Magazine* (Greenwich, Conn.: Management Publishing Co., undated) (pamphlet).

"The procedure: Drop a line to us or make a call to any of the editors of ACH&V to discuss the possibility of an article on the subject you have in mind. If there is interest in your idea make a brief—not too detailed—outline of the subject matter. The purpose of this is to give the editor an idea of what you intend to cover. It gives you a chance to benefit from his experience, for he may be able to suggest improvements." [3]

The paragraph goes on to state, "Next, put the idea on paper," and asserts that this should be easy, too.

Gulf Publishing Company of Houston has an author handbook for each of its publications. Largely similar, they go more into general journalistic matters than some of the advice presentations do, and they contain useful information on the magazines' audiences and criteria for article selection.

For example, each article idea considered by *World Oil* is subjected to a two-part reader-interest test. The first part is a brief statement on what information will be presented in the proposed article that will make it of interest and value, and therefore worthy of publication.

"(The second part) requires that each article be judged on the basis of how many of the following four basic questions can be answered in the affirmative:

1. Will this article help to improve the operations of an oil or gas company, individual operator, or contractor?
2. Will this article help a company, individual operator, or contractor to arrive at a decision or formulate policy?
3. Will this article help a substantial group of subscribers improve personal position and chances of promotion?
4. Will this article carry enough general interest to cause a large number of subscribers (at least half) to want to read it purely from an interest standpoint?

"Each article must answer at least one of these questions in the affirmative to warrant consideration for publication." [4]

In this manner one publication illustrates its criteria for article acceptability. Substitute in these criteria the words that describe your prospective audience and you'll have a good preliminary test for the acceptability of your idea. (If your idea doesn't qualify, can you modify it to make it fit? Is there another audience—publication—for which your idea *is* suitable?)

[3] *Why You Should Write for Air Conditioning, Heating and Ventilating* (New York: ACH&V) (pamphlet).

[4] *World Oil Author's Handbook* (Houston: Gulf Publishing Co., 1964), p. 13.

## BE SURE TO GET A DEADLINE

Assuming that your idea passes the test and conditional acceptance is forthcoming, you may also expect a factor of time commitment to enter the picture. The editor should give you a deadline, normally a couple of months hence, for having a manuscript in his hands. In fact, if he doesn't give you a deadline, ask him for one, and ask him what issue he's shooting for. If his magazine is a monthly he'll have at least a partially developed schedule 6–8 months ahead, and he will have an idea which one your article will fit into.

Your concern about the tentative scheduling of your article goes beyond simple interest and curiosity. You don't want it buried in any magazine's backlog. One advantage of magazines over professional journals is publication speed, so why lose that advantage? Get a tentative publication date, and then do your part by satisfying the deadline for your copy. (Don't relax because there are 2–4 months between deadline and issue date; editor and printer need this time.)

### You'll Need to See the Edited Version

One other assurance is desirable at this time, and that is the assurance that you will see and be permitted to correct the edited version of your article. Most publications of the class treated in this book accord this privilege as a matter of course and in the interest of accuracy. Many of them state their policy in communications to writers. Practically all will welcome your request for the assurance and allow sufficient time for you to examine the copy.

It will be realized that although the title of this chapter is "Selling Your Article," it is concerned almost entirely with "preselling." Your completed manuscript and illustrations are necessary to clinch the acceptance (sale), and it is to the production of these essentials that the next two chapters are devoted.

# 8

# TEXT PREPARATION

The stamped envelope that you mail to the editor will contain both text and illustration materials. Illustrations will be covered in the next chapter; this one is about text.

The text, or manuscript, as it leaves your hands, will be typed at least double-spaced on numbered sheets of unruled $8\frac{1}{2} \times 11$ inch paper. The paper should not be tissue, because the editor inevitably will mark on it, but any quality of opaque uncoated paper is acceptable. Your text will begin a third to a half the way down the first sheet, and your name will appear at the top of each sheet.

## BY-LINES AND HEADLINES

The top portion of the first sheet will present your by-line and possibly a suggested title for the article. Your by-line will be on your article because you have written about a subject on which you are an authority; the consequent effect of your by-line is to give your article, and therefore the magazine, maximum credibility, prestige, and readership.

Headline writing is the editor's prerogative, and he likes to have room to write it on the manuscript. Of course, if you're particularly fond of a title you've composed, you may make a pitch for it and the editor may agree. In one instance an author, chagrined that the magazine didn't use his headline, managed to get it published alone full-size in a subsequent issue. The magazine even stated the author's wish that the "correct" headline be clipped and affixed to the reader's copy of the article. That author, however, was a persistent and persuasive engineer who was unfettered by knowledge of journalistic tradition.

## HOW TO START

How do you start your article? First, write down as briefly as possible what you're going to write about. Name it, don't describe it; at this point you're writing to yourself and you already know what it looks like, feels like, sounds like, and even smells and tastes like.

### Visualize Your Audience

Now define your audience—in writing unless it is quite familiar to you. Presumably you know which publication you're writing for because you have obtained an expression of interest from an editor, as advised in Chapter 7. At least you have in mind a publication that you consider a reasonable target. Study its circulation statement; decide which classifications of readers will want to read your article. Picture those people in your mind's eye: their occupational patterns, information needs, reading habits, temperaments, and prejudices; their points in common and their differences.

### Write Down the Benefits

The next step should be written for sure, because it will yield words for your draft. Put down all the benefits that any individual in your audience might gain from your writing. Consider both the common and divergent interests in the audience and systematically ask, "Will my subject make them (him) richer, wiser, healthier, happier? Will it save him time? Will it help him do his job better? How will it do these good things?"

The procedure to this point may sound like a duplication of steps you took in reaching a decision to write. The route to actual writing is similar to the selling process, but preparations for writing are much more exacting. Now you're preparing to address the reader, first convincing him that what you have to say will be worth his while, then telling him everything necessary to fulfill your promise. With the editor it was possible to be brief and sketchy, and to converse or correspond on points needing elaboration. When you write for publication, it's a monologue.

Pursue possible benefits tenaciously, because they are strong links in the chain between you and the reader. Without them you risk writing for yourself. (Save time; you can talk a lot faster than you can write).

### Canvass for Illustrations

Last, list known and available illustrations that relate to your subject; specify illustrations that could be useful if they existed. Try to figure out which, if any, of them could substitute for or be valuable adjuncts to your words. Write the conclusions of your analysis. By taking this step you will avoid giving the impression that illustrations are afterthoughts, which is just what they will be if you write first and consider illustrations later.

## PICK YOUR STORY LINE

You now have assembled first-rate tools for drafting your article. You'll need one more helper, an outline, and before you make it you'll have to decide on a method of presentation.

### Inverted Pyramid versus Suspended Interest

In the first place, there is a theoretical choice between the journalist's inverted pyramid and his suspended-interest story; the inverted pyramid tells the most important facts in the beginning, then supplies details in a descending order of importance; suspended interest promises something interesting, sets the stage, and finally comes to the point.

The leading-sentence summary of the inverted pyramid contains at least four of the five elements of a news story (the 5 W's): Who, What, Where, When, and Why. An example: Ed Smith built a rocket engine in his garage last winter to drive an experimental drag racer.

Newspaper human-interest stories frequently present suspended interest in its classical form. Most writers can manage the form for a few hundred words, but it takes a craftsman to hold a reader all the way to the end of a feature-length article.

### Suspended Interest Is Secondary

Fortunately for most technical writers and editors (the noncraftsmen, that is), the pure suspended-interest story has little or no place in their field because the "business" reader won't take the time to read a whole article to find out whether it will do him any good. Suspended interest is a useful concept, however, because it can be applied in the first paragraph or so of a technical article to grasp the right reader's attention. A simple example:

> Com Estibles, Inc., reduced packaging costs one-third in its Food City plant, thanks to a system that is available to anyone. Not only does the system save money; it also requires practically no manpower training or re-orientation of supervisory forces.

At this point the writer may summarize what the system is, or he may tell some more of its benefits before delineating the inverted pyramid's broad beginning.

The system was developed for dry-food packaging, but it can be adopted by any industry for retail sales of lightweight items requiring air and moisture protection.

You can see that the reader is being asked to decide whether the article is for him, even before he is given any idea of what the system is. He's saying to himself, "What is it?" Now tell him.

## Inverted Pyramid Prevails

The "what" of the technical article is quite likely to be more useful and important to the storytelling than any of the other journalistic "W's" except the "why." The "why" includes the benefits you so painstakingly wrote down; it may also include some other reasons for your subject—reasons that were compelling for its development and which contribute essential background information to the article.

The "who" gives authenticity to the story, gains recognition for you, and enhances the image of your company (you and your company are the "who"). The "when" is of variable value. If newness and speed of accomplishment are important story elements, the "when" is important; if you are presenting ideas long held but now revealed, the "when" may be insignificant. "Where" for a project or device satisfies a reader's need for location and orientation. Some writers deliberately withhold place names for a few paragraphs, feeling that many readers will reject a story involving a place with which they do not identify. These readers should be hooked, so the theory goes, by benefits and outstanding features before they have a chance to exercise their geographic prejudices. This approach may be valid for many readers and it won't hurt the ones who like to know where they are; they may even be charmed by the moment of suspense.

## Opening Paragraphs Command Attention

The subject of opening paragraphs is well-nigh inexhaustible. H. J. Tichy, in her book on writing, emphasizes the opening by giving the topic an entire chapter and calling it "Two Dozen Ways to Begin."[1] Editors know that the golden goal of high readership depends strongly on the arresting quality of these paragraphs. And therein lies a comfort for the author who despairs over his opening. The editor is almost certain to rewrite the opening. At least he will give much thought to the possibility of improving it and he won't begrudge the time it takes.

[1] H. J. Tichy, *Effective Writing for Engineers, Managers, Scientists* (New York: Wiley, 1966), p. 25.

## Plan Before You Make Your Outline

The editor wants the rest of your text to be written competently, however. If it is poorly organized, contains chaotic sentences, contradicts itself or seems incomplete, he is going to approach the editing job reluctantly.

Organization, to be good, must *precede* actual outlining. Select an approach.

According to J. H. Mitchell, " 'Standard' approaches are chronological, logical, physical-spatial, or fixed-formula, but combinations are always possible and are often preferable." [2] Professor Mitchell's ensuing discussion presents the first three approaches as techniques to be applied where appropriate within "organization by formula." This emphasis is proper for his audience, the writers for journals, because many of these publications specify components and sequence for complete and acceptable papers. Formula writing drops out of the picture for magazine writers, however, leaving the logical, physical-spatial, and chronological approaches in readiness to serve. Logic, the tool of argument and persuasion, is useful for the expression of opinion, but it is at best a subordinate technique for technical articles.

## Description and/or Chronology

Your organizational choice, then, will be physical-spatial or chronological, or a combination of the two. If you're writing about an object, be it an earth dam or a gold watch, describe it (physical-spatial). State its function and name its components; then describe the components. If your story is about a component, say so first; then place the component in functional and physical perspective with related components; finally, go into your detailed description of the component.

Compare the object with the object it supplants or surpasses, pointing out superiorities of size, first cost, performance, ease of maintenance, or other pertinent features, and noting any offsetting disadvantages. Less tangible subjects, such as management methods and theories, also are susceptible to classified description and analysis.

Description may take a back seat to chronology where there's action, as in a new process, an unusually fast construction job, or a fascinating story of events leading to a breakthrough. Chronology is easier to manage where simple sequences are involved. Where there are concurrent events and sequences, a combination of classification and chronology is necessary.

[2] J. H. Mitchell, *Writing for Technical and Professional Journals* (New York: Wiley, 1968), p. 73.

## NOW WRITE YOUR OUTLINE

When you have decided on the order of presentation that best suits your subject, write an outline. Start with three pages headed, respectively, "Opening," "The Story," and "Closing." On the "Opening" sheet, follow through your planning by jotting down key words for basic facts and reader benefits. Under "The Story," name your main divisions, leaving plenty of room for subdivisions; you may need more than one sheet. The "Closing" sheet may end up blank. Magazine articles usually are devoid of formal conclusions or final summarizing statements such as those found in theses, learned papers, and reports. In traditional practice, however, the last paragraph is a catch-all for credits. An example:

Laboratory investigations leading to development of the new process were the responsibility of Dr. John Lovelace, Chief Chemist of XYZ Corporation. The author was in charge of pilot-plant investigations, and handled all liaison with ABC Engineers, Inc., designer-constructors of the plant.

There is some question whether this practice of lumping credits at the end is not more archaic than traditional. It is consistent with "old" technical writing style, which demands that everything be in the third person, and that the passive voice prevail. This allegedly dignified style brooks no "I did it," or "He built it," so both he and I must sit modestly in the last paragraph and hope the reader will struggle all the way through to find us. (Thank goodness for my by-line.)

Thank goodness, too, for the concept of readership, which has weaned writers and editors away from old styleways and forced them to put life and punch into their copy.

You can build both life and punch into your article from the start by keeping two precepts in mind as you complete the story outline.

*For life*—Write in the active voice—first, second, *or* third person—wherever possible, without stretching the bounds of natural expression and modesty.

*For punch*—Omit details that do not strengthen your main points.

Both precepts are still more important in drafting the text than in the preparation of the outline, but the outline is the device that will guide your thinking along these positive lines.

### For Life, the Active Voice

To assure maximum use of the active voice, write complete sentences wherever you can in your outline. For example, where you are tempted to write "forms" to indicate a step in concrete construction, write, "Carpenters erect forms." This way you'll forestall the passive, "Forms were erected," which just doesn't gain the reader appeal that the active voice does.

Save outline-writing time by devising a code for outline-sentence subjects. Select as subjects the persons and devices that act, not the objects that are *acted upon.* The code, of course, should include each entity that you plan to save from last-paragraph oblivion.

Note that the complete-sentence approach requires you to select verbs. This, too, is good, because it comes at a time when your thinking is not clogged with details and intricacies of style. Make the verbs work for you. Choose verbs that express action. Action verbs in active voice spur lively reading. Moreover, putting verbs in the outline gives you a second chance at the best choice when you get to the draft.

## For Punch, Eliminate

As to elimination of extraneous detail I do mean elimination, not avoidance. With your main framework of description and chronology established, you should write everything that seems pertinent under each heading. That's the reason for starting with plenty of paper. Then examine each item. Ask yourself whether it supports your main points; cross it out if it doesn't. This is elimination, deletion of material you've written. Don't grieve; you have done your reader a favor.

## Reappraise Sequence

Consider the sequence of the surviving items; where revision will improve the story line, mark the changes. Look for points that belong together; assemble them.

## Flag Illustration References

Make big solid circles around notes that pertain to illustrations. These notes are special guideposts for your writing because they tell you where you'll have pictorial help for your wordsmithing. If you intend to present tabular material, it, too, should be flagged at the appropriate reference points.

You now have an outline that is a real working tool. Probably you can draft from it as it is, but if it looks a mess there's nothing wrong with recasting it.

## DRAFT YOUR ARTICLE

Handwrite, type, or dictate your next draft, whichever method suits your inclination, temperament, and experience. Handwriting pleases people who like to fiddle with their style as they go along and who have the patience to wait for their fingers to catch up with their thoughts. Dictating is for authors who either don't have this patience or do have a proficiency at speaking from notes. (Typing, for those who can type, is an in-between method.) The impatient dictator may be impatient with his outline, too, a fact that makes it doubly important for him to recognize his weakness and slow down, at least to the line-by-line speed of his outline. As an extra precaution, however, he should specify triple spacing for his transcription.

Whatever your method of making your draft, it should end up typed —double-spaced or triple-spaced, depending on the amount of rewriting you think it will need.

Now draft the trimmings: captions for illustrations, tables, and any other written material that will be typeset separately from the text. Each kind of material goes on a separate sheet of paper.

### Captions Go on a Separate Sheet

Write "Captions" at the top of one sheet, then write them as a series of double-spaced paragraphs separated by triple or quadruple spaces. Put an identifying symbol to the left of each caption so there will be no question about which illustration it describes.

Captions should start big, stating each feature of the illustration that you want the reader to see and telling how it relates to the story. If identification of people is at all pertinent, give full names and titles. Do not, however, succumb to any temptation to tell what's *not* in the illustration.

### How to Handle Tables

Each table should be typed separately. Double spacing is best, even though it may run copy to more sheets. If the normal margins of the sheet don't accommodate all your columns, turn it sidewise. Still wider tables should go on oversize sheets. Neat hand-drafted tables are acceptable, but use capital and lower-case lettering, as you would with a typewriter. It is better to furnish a full-width table, no matter the size, than to present adjacent pieces that must be aligned by somebody else, thus increasing opportunity for error.

There are practical limits to table widths, of course, because magazines

## Contractor Failures by Type . . .

| | Number 1968 | Number 1967 | % Chg | Liabilities (000) 1968 | Liabilities (000) 1967 | % Chg |
|---|---|---|---|---|---|---|
| U.S. Total .................. | 1670 | 2,261 | —26 | $212.5 | 323.7 | —34 |
| General building contractors..... | 656 | 867 | —24 | 135.3 | 238.9 | —43 |
| Building subcontractors ........ | 903 | 1,243 | —27 | 58.2 | 71.4 | —18 |
| Heavy and highway contractors.. | 111 | 151 | —26 | 18.9 | 13.4 | +41 |

## . . . and Why They Failed*

**NEGLECT**—48 firms
*Apparent Causes*—Bad habits, 17; Poor health 15; Marital difficulties 7; Other 9

**FRAUD**—12 firms
*Apparent Causes*—False financial statement 3; Irregular disposal of assets 8; Other 1

**LACK OF EXPERIENCE OR INCOMPETENCE**— 1,517 firms
*Apparent Causes*—Inadequate sales, 421; Heavy operating expenses 441; Receivables difficulties 244; Inventory difficulties 21; Excessive fixed assets 64; Poor location 10; Competitive weakness 398; other 90.

**DISASTER**
*Apparent causes*—Fire 4; Flood 1; Burglary 1; Employees' fraud 1; Other 5.

**REASONS UNKNOWN**—81 firms

** Because some failures are attributed to a combination of apparent causes, these totals may exceed totals of general underlying causes.*

*Source—Dun & Bradstreet*

**Figure 2. Same-size facsimile of table printed two columns wide shows use of various type sizes and faces to indicate relative importance of data elements. Fonts range from 12-point bold for headings to 6-point italic for the footnote. (Courtesy of Engineering News-Record.)**

rarely will print fold-outs. For example, on a 7-inch printed-area width, a line of 6-point tabular material would have a maximum character-and-space count of about 150. Figure 2 shows the use of 6-point and larger types on two columns. Few magazines would be willing to go as small as 6-point for a three-column table; 8 point would be more reasonable, with a high count of 125. And don't ask any magazine to print a table sidewise on the page; readership theory holds that all material should be intelligible with the page held in its normal position. (A table typewritten sidewise on a manuscript sheet can be set in acceptable type size within ordinary magazine page margins.)

Every table presents one significant problem of choice to the writer: which characteristics to show in columnar arrangement and which to arrange horizontally. Melba W. Murray gives these helpful rules: [3]

[3] M. W. Murray, *Engineered Report Writing* (Tulsa: Petroleum Publishing Co., 1964), p. 60.

"1. State the problem of the comparison in words that will show (a) what you are ultimately comparing and (b) what characteristics, properties, and other attributes you will be using to make this comparison.

"2. Arrange headings so that major classifications for ultimate comparison and numerical values to be compared will appear in vertical columns—not in rows.

"3. Place all subheadings (all 'modifiers,' in effect) under the major headings they describe."

A simple example follows the rules:

**Characteristics**

| *Rock* | Porosity (%) | Permeability (*md*) | Grain Size (*mm*) |
|---|---|---|---|
| A | 1 | 4.5 | 0.02 |
| B | 2 | 5.5 | 0.04 |
| C | 3 | 6.5 | 0.08 |

## Would Graphics Be Better?

As a matter of fact, you're probably risking lowered readership by submitting any but the simplest, most readily understood tables. If there is any way to present the material graphically, do it. Curves, bar charts, pie diagrams, and pictographs have eye appeal and can summarize lots of data. Consider, too, the desirability of presenting both graphics and complete tables, where you know that many readers will want your detailed data.

## Abstracts and Descriptors

Other written materials that might be expected of you are an abstract or summary, and key words or descriptors.

Some of the more technically oriented commercial magazines have adopted the practice, prevalent among learned and scientific journals, of printing abstracts and descriptors for use in data-retrieval systems. They specify maximum wordage to fit on a 3 × 5 inch index card, and they issue instructions on the nature of the abstract they want: whether it should be informative (tell premises and conclusions) or indicative (list topics).

Descriptors are supposed to permit access to articles through data-processing systems. For example, an article about machinery and controls for a twin-leaf bascule bridge across the Chicago River might be tagged tentatively with the following descriptors: "bridges, movable"; "machinery"; and "controls." These descriptors would turn out to be the right ones if they were in the vocabulary of the retrieval system being used. A check against that vocabulary list would reveal discrepancies and dictate corrections. Subsequent input of the descriptors with the article's code would permit retrieval by anyone seeking data on any or all of the given descriptors.[4]

## REVIEW FOR COMPLETENESS

When you have finished your draft you will feel a great sense of accomplishment—and you will be right because the hard part will be over. But you still have some detail work to do to assure completeness and coherence in the finished text.

Make a reminder list, like this:

Outline
Illustration references
Quote references
Credits

Now lay your new double-spaced draft beside the outline. Scan the draft, noting its points and checking them off in the outline. If you've missed anything and you still need it, write it in—between the lines if there's room, or the other side or on an attachment. If you've written about points not in the outline, and you decide they're nonessential, cross them out.

Check next against your illustration list. Be sure your text contains a reference to each one, and that the reference makes it clear which illustration is meant. Sometimes editors find it difficult to make the connection. Ironically, but for good editorial and readership reasons, the editor may delete some textual references to illustrations. That's his business.

Quote references, too, may be informalized or deleted by the editor, but you should check to make sure they're all in the text when you're through with it. Depending on your feel for the publication, you may put the references with the quoted material, with the aid of word devices such as "According to Dr. H. T. Smith of the Newton Observatory . . . ," without supplementary footnotes; or you may do it all with footnotes.

Credits have a way of getting limited to the author's own organization and that of his client (if any). True, they *can* get absurdly numerous, but you ought to be fair; at least you should consider which responsible parties would dislike being left out, particularly if they can do you some good in the future.

[4] Several technical societies publish "microthesauri" of descriptors that are accepted within their disciplines or subdisciplines. At a higher level, *Thesaurus of Engineering and Scientific Terms* (Washington, D.C.: Engineers Joint Council—Department of Defense, 1967) is a standard work.

Anyway, make a list if you didn't do it at the end of your outline; then check off the ones you mentioned in the text draft and decide what you'll do with the rest. You may end up with a final credit paragraph after all.

The checklist phase of text preparation puts the finishing touches on the first draft. Most editors would be pleased to have a manuscript that's even as carefully prepared as this first draft. But you should go the whole way.

## REVISE FOR THE READER

Put yourself in your reader's frame of mind. Try to read the draft as if it were all new to you. Try to find sentences and paragraphs that don't satisfy you: descriptions that don't fit together; discontinuous narrative; incomplete logic; extraneous material. Have you asked questions that you haven't answered? Don't stop to correct these shortcomings, don't let them slow you down any more than they would any reader (but keep going; he would be likely to quit after stumbling just a few times.) Just make check marks in the margin and read on.

As a back-up step, you may ask a fellow worker to become your "audience" and check items he'd like clarified. Furthermore, this point in progress could well be the place for clearance review by your superiors; most organizations require management approval of articles written by employees.

Now, having gone through this exercise in reader thinking, you become the author again. Each check mark flags a revision task. Hopefully each task is small: a word added or deleted, phrases transposed to give better sequences; a change of voice. However, there's nothing unusual about complete rewriting of paragraphs or whole pages at this stage. The saving grace of revision is that you do it while you're in the swing of writing, you're full of your facts, and you've already eliminated one way of writing the material you've flagged.

The question, "Have you asked questions that you haven't answered?" is largely figurative. A literal illustration would be a sentence like, "And what were the indirect results of this experiment?" followed by text that in no way answered the question—by neither presenting the indirect results nor stating that they were unknown. However, the quoted question is a typical device of suspended interest, which already has been deprecated as a technique for technical articles. What you're more likely to write is "Experimental results were both direct and indirect." Then, if you tell direct results and ignore the indirect results, you have in effect "asked questions" that you haven't answered. Another way of putting it is that you have promised information that you have not produced. If you can't produce the information, or if you don't deem it essential to the story, don't promise it; if you aren't going to answer a question, don't ask it.

This little exercise in redundancy may seem like overstressing the obvious, but many professional writers have never lost a horror of violating its principle. "Leave no questions unanswered" is likely to be one of the first pearls a pro presents to a neophyte.

Correcting the items you've flagged should develop your article to the point where it is technically sound.

## REVISE FOR THE EDITORS

If you have more time, or if you are inclined toward craftsmanship in such things, your article should now receive the finishing touches that bring joy into an editor's life.

Now is the time to make sure each word is the best one, each sentence complete, each paragraph a cohesive unit. Now is the time to apply the principles and tests of books on plain talk and fog-free writing; to hope that you've somehow imparted simple grace and rhythm to your text. Style, too, deserves attention at this time, to assure uniformity in capitalization, word compounding, punctuation, numeralization, and abbreviation. A style guide is presented in Appendix A.

### The "Wrap-up"

Final typing must be at least double-spaced. The 8½ × 11 inch paper size is preferred, but 8 × 10½ inches is all right. Avoid legal size, however. Your name and address can go on the upper left corner of the first sheet; page numbers upper right. Put an identifying title a few spaces down, then drop to the middle of the page to start the text. Subsequent pages require identification (your name or a word that relates to the subject) at the upper left and consecutive numbering upper right. Text, captions, 8½ × 11 inch tables, and other written materials follow with the same top-of-page designations. Outsize tables will be clipped together with illustrations in the envelope you send to the magazine.

Your last step in manuscript preparation is proofreading the final typescript. Follow standard copy-marking practice (Appendix B), using a pencil rather than a pen. The fewer the corrections the better, of course, because editors like to go to work on clean copy, but they don't require letter-perfect copy.

# 9

# ILLUSTRATIONS

A picture may be worth a thousand words, but illustrations for technical articles are likely to be afterthoughts, affixed to the manuscript when the struggle with words is done.

Even the order of topic presentation in this book betrays the tendency to relegate illustrations. We covered text in Chapter 8; now here's Chapter 9, "Illustrations." The logic of this sequence probably is correct, but for many technical people it cloaks a paradox. These people, accustomed to thinking first in graphics and mathematics, are asked to write words first, then add pictures. Actually, illustrations are indispensable elements of many if not most technical articles, and the final treatment of illustrations influences the final edit of the manuscript.

This chapter has two principal parts. The first tells the functions of illustrations, their types, and how to prepare them for the editor and printer. The second part covers photography—how to obtain the right photograph.

## THE FUNCTIONS OF ILLUSTRATIONS

Functions of illustrations are to supplement and clarify text, to conserve space, and to attract readership. The writer is concerned most with the first function; the publication does its best to conserve space and to assure readership.

### Illustrations Supplement Text

Whenever you get involved in a difficult description situation, ask yourself, "How would an illustration help me?" Chances are that the illustration already exists; if it doesn't, you can make a sketch or a graph. Don't worry about the quality; that can come later. The illustration will help you complete the word description, and you'll be off and running to make your next point.

## Space Conservation

A little arithmetic will demonstrate the space-saving property of illustrations, assuming that the picture is, in fact, worth a thousand words. A thousand words takes up about five-sixths of the typical 8½ × 11 inch magazine page. Few illustrations end up at full-page size in technical and business magazines; most are sized somewhere in the one-fourth to one-half page range. It is conceivable in the absurd extreme, therefore, that good illustrations can reduce the space needed for your story by nearly two-thirds!

From the publisher's standpoint, space conservation normally is an end in itself, because he strives to pack as much good editorial material as possible into the pages budgeted for this purpose. He also likes to keep the article as short as possible, consistent with his concept of service to the field, because he knows that brevity enhances readership. Studies continue to confirm that the optimum article length from the readership standpoint (percentage of surveyed readers who remember reading given articles) is about two pages.

## Eye Appeal

A secondary use of illustrations, fully as important as space conservation from the readership standpoint, is eye appeal. A photograph of a handsome building or a blown-up detail (however incomprehensible in itself), indeed any illustration that will arrest the reader's attention and make him look to the text, has eye appeal.

Good sense dictates that the attractive illustration relate to the article's topic, to avoid confusing or amusing the reader at the publication's expense. On the other hand, if the eye-appeal illustration also happens to be highly informative and pertinent to the topic, so much the better.

The editor is the judge of eye appeal; in fact, he is the one who decides which illustrations will be used and how they will be used. If you have two that illustrate the same point, send both; if you have more, send them all. He likes a selection, and his fresh, independent viewpoint will discern journalistic values that assure the appropriate choice.

## TYPES OF ILLUSTRATIONS; HOW MAGAZINES USE THEM

Despite constant improvement of graphic-communication effectiveness, the choice of illustration types remains the following: drawings based on existing work, graphs and diagrams, photographs, freehand art (including cartoons), and composites. Each of these types is a tool the editor chooses and manipulates to the best advantage of the article he edits.

### Line Drawings Based on Existing Work

The editor's experience with illustrations is particularly useful to the writer who needs to present line work with his article. An author often has fragments of the ideas to be portrayed on two or three different sheets in a roll of large prints. Bringing the fragments together and simplifying them into a readily understood line drawing is a function of the writing-editing process.

The least the writer should do to initiate this process is to submit with his manuscript the basic drawing or drawings with the pertinent material encircled. The most he could do would be to prepare an inked linen tracing using the magazine's drafting and lettering style and including exactly the amount of detail that the editor would choose. However, this completeness is more than any adequately edited publication needs.

Editors have illustration talents available to help them complete the line-drawing ideas indicated by authors. A large publication will have its own illustration department; a small one may hire free-lance draftsmen and artists to do the work. In either case the process is the same, comprising the following steps.

Instructions to illustrator
Drafting by illustrator
Corrections, if required
Instructions to printer

Taking the somewhat complicated example suggested above, in which a line illustration is to be derived from three different sheets of a project's drawings, here's what happens. You mark the areas of the drawings that pertain to your story. Yellow crayon is good on blueprints; red on dark-line prints. You may also indicate by a separate sketch the positional relationship desired in the final drawing. Encircle dimensions and other legend you want retained and run wavy lines along lines that are not essential to understanding. Important lines may be strengthened with the marker, a note being appended suggesting more line width in the final drawing.

The size at which the drawing will be published is marked on the layout sketch, which is given, together with the three marked drawings, to the illustrator. He first photostats the designated areas of the drawings, reducing or enlarging each to a common scale large enough to permit easy drafting. A multiple, 2 or 3, of the size to be published is normal for illustration drafting, but any other convenient factor will do.

Whatever the enlargement factor, lines must be wide enough and lettering must be big enough to retain legibility after reduction by the factor's reciprocal. The smallest legible letter is about $\frac{1}{15}$ inch high. If drafting is done at 3 times publishing size, minimum lettering size will be $\frac{1}{5}$ inch, definitely within the range of normal draftsman capability.

Although drafting convenience is the principal advantage of enlarge-

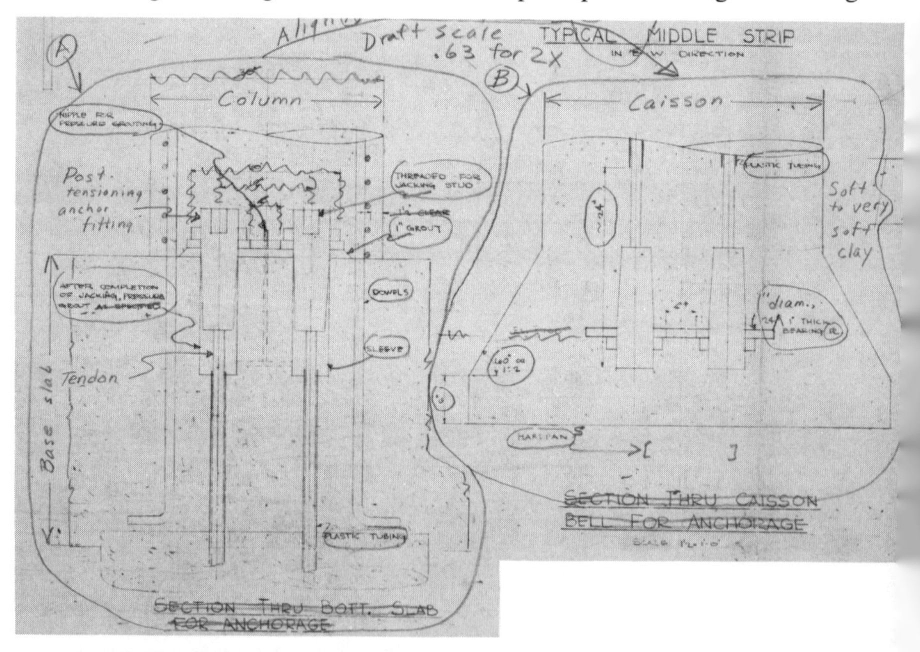


**Figure 3. How a designer's drawing becomes a line illustration. Marked-up drawing fragment (shown in linecut above, reduced to about one-third of drafted size) was reduced by a factor of 0.63 to twice the planned one-column cut size. Illustrator then traced prescribed line work, added symbols, and impressed type-set lettering. Half-reduction of illustrator's drawing took place in printing process, the result being the line illustration on p. 85, presented at the same size as originally printed in Civil Engineering, December 1967. (Drawing courtesy of Albert Kahn Associated Architects and Engineers, Detroit.)**

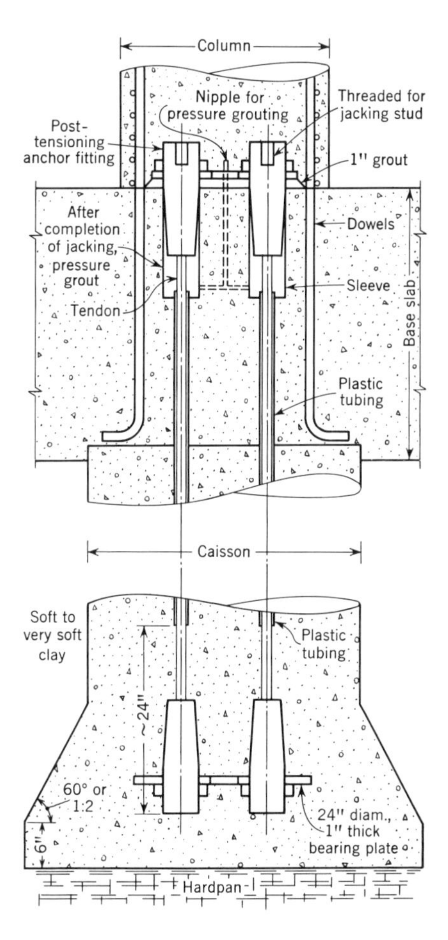


ment, there is another benefit: reduction to publishing size makes minor imperfections of drafting technique less visible to the reader.

Drafting technique, or at least drafting style, is standardized by most publications, so that within a given publication the line illustrations present a uniform effect. Points of style include borders (most magazines use them; some don't), line widths, arrows on dimension lines, lettering, and methods of providing shading.

In recent years the options available for lettering have increased, a popular method being the application of adhesive characters. Produced by several commercial sources, the characters are available in a great variety of type faces and sizes. Each font, comprising letters, numerals, and symbols in various quantities, depending on normal frequency of usage, is printed on a transparent (acetate) sheet with a nonadhesive backing sheet. Characters to be applied to the drawing are cut out of the font sheet and pressed into place.

Another lettering option is typography. In this approach only the line work is done by hand. Legends, notes, and dimensional data are typeset in the final reproduction size; proofs of high quality, called reproduction proofs, are clipped to the drawing for "stripping in" on the reproduction medium.

A few publications continue to use freehand lettering for their line illustrations, somehow preserving the old-fashioned drafting-room craftsmanship, but most hand lettering today utilizes guide devices such as those with the trade names of Wrico or Leroy.

Whatever the method of inscription, the line illustrations must be checked. Correction instructions are marked in blue, not because the blue pencil is the editor's traditional tool, but because blue doesn't reproduce under the engraver's camera. Thus the instructions don't have to be erased when corrections are made, and, when the drawing comes back from the draftsman, the editor can see his correction marks to recheck the work.

One more step may be handled by the draftsman. If shading is required, the areas to be shaded are marked on the corrected drawing. The draftsman then affixes pressure-sensitive media to the designated areas. The process is the same as for adhesive lettering, the shadings being available in a variety of cross-hatchings, dots, etc., sized for several reduction factors.

Shading also may be accomplished in the printing process. Where letterpress is to be used, dot-type shadings are available in several "tints," or percentages of blackness. Called Ben Day, the tints are made during the photoengraving of line cuts.

The instruction to the printer for Ben Day is the same as the instruction to the draftsman for affixing the prepared shading media. Other instructions to the printer are simple, the most important being the size at which the drawing is to be printed. One dimension is sufficient. Usually it is the horizontal dimension, and it is likely to be standard for the publication's style: single-column width, double-column width, or other size.

The normal line illustration is printed in black (or another color) on white. If a reverse is desired—white lines on black background—this deviation is specified.

For printing in more than one color, the lines and lettering for each color are encircled and designated by a code number, letter, or the name of the color, so that separate cuts or negatives can be made for the respective imprints.

## Graphs and Diagrams

Graphs and diagrams are treated much the same as drawings based on existing work, except that they are more likely to be originated especially for the article. Statistical data that would make tiresome reading, and which would require study if tabulated, usually can be presented graphically so that meaning can be grasped at a glance.

A variety of forms are available: curves, bar graphs, pictographs, and pie diagrams, for example. These devices portray trend, quantity, and division. Block and flow diagrams can function to simplify organization, system, movement, and process information.

Because the responsibility for originating appropriate graphs and diagrams centers on the author, some helpful hints on preparation are pertinent. (For an extensive and highly understandable treatment of the topic, see *Graphic Communication,* by William J. Bowman.) [1]

Probably you will plot curves and bar graphs on standard cross-ruled sheets, purchased from a stationer or drafting-room supply house. Whether they be arithmetic or logarithmic, the sheets will have more lines on them than you will need for adequate portrayal of your information. For example, a grid of 1-inch-spaced lines may be sufficient for reading purposes, whereas you will need tenths of inches for plotting. Instructions to the draftsman, therefore, will tell him which cross-ruled lines to draw, in this case the 1-inch-spaced ones. They're guide lines, of course, so the lightest line should be used.

This simplification of curves and graphs is essential for black-and-white printing. Where color is available, there exists an attractive alternative: printing all the cross-ruling in a light color, such as the blue or orange in which it usually is printed by its producer. There is no need for the draftsman to redraw the grid lines. He can do the ink work on a fresh sheet of the same paper you used; the separation will be made in the reproduction process, with the help of appropriate filters or films.

Pictographs can enhance eye appeal where comparative statistics are presented for single physical objects, such as numbers of houses, cars, or people. One silhouette drawing of the object is reduced to various sizes, representing the numbers to be compared. To play fair, you should make the areas, rather than single dimensions, proportionate; thus the square root of a multiplying factor would be applied to a single dimension to get the linear relationship. There is no need to carry the fair-play doctrine to its logical extreme by insisting on three-dimensional proportions. The

[1] W. J. Bowman, *Graphic Communication* (New York: Wiley, 1968).

paper has only two dimensions, and besides, pictographs aren't that real. Imagine a million people being one person 1 inch high and four million people another person 2 inches high.

Logic, rather than scale, prevails in the approach to block and flow diagram preparation. Thus your initial rough sketch will be sufficient for submission to magazines that have normal art services at their command. The vocabulary is simple: geometric shapes (blocks, rectangles, circles, etc.) to indicate entities; lines to indicate connections; arrowheads on lines or relative position among blocks to indicate flow or hierarchy, respectively. A few symbols, such as plus and minus signs, may be necessary for more complex or abstract situations, to indicate alternatives or optimizations.

## Photographs

Turning from the abstract to the concrete, for realism in journalism the photograph is the standby. If it exists and if it can be found, it is the most convenient of the illustration forms. If it's not at hand, but nevertheless is deemed essential to the story, the photograph can be a will-o'-the-wisp.

Because photo acquisition is largely the author's responsibility there will be more on this topic in the latter part of this chapter. For now, the assumption is that you have the pictures. You refer to them in your text and you write captions for them. You also mark lightly on their backs some form of identification, such as figure numbers or descriptive words.

If a credit line for the photographer is required, this fact also should be noted on the back.

Editors prefer glossy-finish photos because they present the best detail; some publications even insist on them and specify a minimum size, such as $5 \times 7$ or $8 \times 10$ inches. You should submit enlarged glossies if you can, but semimatté snapshots can be fully satisfactory, providing the negatives were appropriately exposed and the intended details are clearly visible.

Either black-and-white or color prints are acceptable. The latter yield good black-and-white reproduction, and they present the option of color reproduction. Today more and more low-circulation magazines have money in their budgets for color.

Details in some photographs require graphic explanation to the editor if not to the reader. A common technique applied by authors and editors alike is to place notes on a tissue overlay. Paste the edge of a piece of tracing paper to the back of the photo at its top and fold the paper down to cover the front; trim the paper to the photo's size. Now mark lightly, encircling significant details (visible through the paper) and identifying them. Areas you don't consider pertinent can be covered by crosses or marked, "may be omitted." You may even specify a deletion, but it's probably better to use another picture. Retouching is an art that, at best, impairs a photo's air of authenticity; less than best can be grotesque.

Retouching nevertheless is part of the treatment that a photo may receive before publication. The least treatment is sizing and cropping instructions, marked on its glossy surface with a grease pencil. The editor decides which photographs to publish, how big they should be, and what their proportions should be.

If the photo is just right, he marks "crop full" in the narrow white margin, adding the horizontal dimension in inches and fractions. As an alternative to the "crop full" instruction he may make crop marks in the shape of a "V" in the four margins, the apex of each mark touching the edge of the image.

If only a portion of the photo is to be presented, the editor places his crop marks so as to delineate the area. A crop mark in the image has at least a short line crossing its apex, like the bottom line of a handwritten Roman numeral five. The short line is sufficient where the photo is properly

**Figure 4. To rectify a photo that was taken with the camera tilted sidewise, the editor draws an edge-of-image guide line parallel with the vertical lines in the image portion of the photo. This photo of navigation lock construction is marked for 6¾-inch width.**

**Figure 5. Where camera is tilted backward to get a tall subject, verticals converge to the zenith. Crop marks should make edge-of-image guide line parallel with the vertical in the center of the image. Here the over-all photo is framed correctly; cropping for entry-hall exterior detail takes vertical from column in center of area selected.**

squared; that is, where the illustration's edges will be parallel with the photo's edges. Some photos, however, appear to be tilted sidewise—because the camera was tilted sidewise, presumably. Here a full-height edge of-image line should be drawn parallel with a vertical in the picture, to give the printer a strong guide. The other crop marks can be normal, or, better, the rectangle can be completed with full guide lines.

Cropping requires still more attention where vertical lines converge in the photo, usually because the camera was tilted back to get all of a picture, as for a building. Here a vertical line in the middle of the image is chosen as the parallel guide for the side crop lines.

Beyond cropping and sizing, which is all the editorial treatment many good photos need before they go to the printer, there may be art work. Two principal kinds of art work are symbols and retouching. Symbols include spots, patterns, lines, and lettering, intended to locate and explain details in the picture. Black is used on light-toned areas; white on dark. Retouching includes fine-brush (camel's hair) and air-brush work. The fine brush applies tones to mask or camouflage photo defects and details that might confuse the reader, an effort being made by the artist to create an inconspicuous, rather than an artificial, effect. The air brush, a small spray-painting device, applies white-gray-black tones uniformly in any thickness. The thinnest coverage by a light tone imparts a barely perceptible haze to the photo surface. Hazing all but the most important part of a picture subtly draws attention to that part while retaining the visual relationship to its surroundings. This treatment could be applied to an aerial photo of a town, to illustrate an article about its civic center. All but the civic center would be thinly air-brushed. Thick air-brushing obliterates the image. It can provide a neutral panel for explanatory notes, eliminate extraneous or unsightly details, and establish clear-cut skylines or neutral foregrounds. As in other kinds of retouching, it is difficult to avoid the expression of artificality when sharpening skylines. However, this treatment can be useful where the skyline is important and it is difficult to distinguish from leaden skies above. (The air brush does make convincing clouds.)

## Freehand Art

Freehand drawing, painting, and other creative art work commonly found in general magazines do not constitute major elements in technical and business magazine illustration. They have their place, however, primarily in the forms of simple sketches or cartoons and "artist's conceptions"—architectural renderings and images formed in an artist's mind by words he hears or reads, rather than by objects he sees. If the artist is also the author, so much the better.

Occasionally you will find that a simple sketch will help you express an idea; draw it and submit it with your manuscript. In some cases the very spontaneousness of such a drawing gives it a charm and feeling of authenticity that redrawing by a professional artist would lose.

## Composites

The foregoing forms of illustration—technical drawings, charts and graphs, photographs, and freehand art—are essentially separate techniques of illustration. The possibility of combining techniques, however, was touched on in the discussion of photographs, wherein superimposed line work and lettering can indicate, emphasize, and explain image details.

Combinations of techniques, or composites, expand the role of illustrations in efficient communication. Actually there is but one combination, that of line work (solid-color material) and tone work (material, such as photographs, having various shades to be portrayed). But the variety of line and tone forms makes the composite a highly flexible medium.

Line and tone work may occupy separate areas in a composite, or one may be superimposed on the other. An example of separation would be a line map of the United States, surrounded by photos of major dams; an example of superimposition would be the same map printed over a large photo of one of the dams. Charts and graphs portraying air-pollution statistics can be printed against a backdrop photo of a large city's smog pall. A perspective of a proposed building can be fitted into a photo of its site, either by superimposition or by dropping out the photo area outlined by the perspective and inserting it in the vacated area.

## CAPTIONS

The techniques of compositing can make an illustration completely self-explanatory, but for most illustrations a statement called a caption (or cut-line or cut copy, etc.) must be presented. Writing captions is part of your task. One should be written for each illustration you submit, including any that you deem self-explanatory, except that one will do for each set of alternative illustrations. Tell what the illustration is supposed to tell; give full proper names (people, equipment, places); do not hesitate to duplicate material that is in your text.

Captions are written separately from text material, for at least two good reasons: they usually are presented in a different type face from that of the text, so must be set by a separate mechanical operation; and the necessities of article layout (the process of allocating page space to display type, text type, and illustrations) make it impractical to place most illustrations in direct sequence with their pertinent text.

## HOW TO OBTAIN THE RIGHT PHOTO

Furnishing illustrations that are truly essential to clarification of the text frequently requires specific photographic effort. When you start to write, you may believe that you have lots of pictures to help you tell the story, but when you examine them critically you may find that they somehow miss important points you need to depict.

The best way to get the pictures you need is to take them yourself, assuming you don't need aerial photographs or special techniques such as photomicrography. Second best is to accompany a photographer and direct his work, but even this close association embodies hazards of poor communication, and it can be expensive if the photographer is a professional. His pictures may look better than yours would, but unless you looked through his view finder before every shot you're likely to find that he missed details you considered important.

### Send a Photographer

When time or distance prevents your presence, of course you must rely on instructions to a photographer. Write them. You can't be too specific. One or more sketches are needed for each photo, showing the arrangement of the objects and a plan view specifying camera position. If the subject is outdoors, specify the time of day for the right lighting conditions.

### Better: Take the Photo Yourself

Such a shooting script will be useful to your own photography, too, saving you time and film in the field.

"The field" may be literally out where a construction job is under way or a completed work is in operation. Or it may be in your office or laboratory. Wherever the object is, you will need the right equipment and more time than you thought it would take.

*Equipment.* Depending on the subject, the right equipment may be a fixed-exposure box camera; or it might have to comprise a widely adjustable camera, a variety of lenses and view finders, artificial lighting facilities, a close-up device, tripod, and exposure meter.

*Camera Choices.* For the casual photographer, the self-printing film camera (Polaroid Land) is useful in many circumstances, its principal advantage being that results become known immediately. Exposures and poses can be corrected on the spot, and there is no waiting for processing as there is for regular film. The film cost can be significantly higher, however, and this is one factor that influences more experienced camera users to stick with negative-film equipment. The cost advantage is largely offset, however, by the tendency of negative-film users to shoot more pictures. The absence of on-the-spot waiting for film processing permits quicker reshooting.

**Figure 6. How a wide-angle lens solves a common photographic problem. Cameraman can come closer to subject of given width, thus eliminating distracting foreground details. An elevated viewing location helps, too.**

For most article-illustrating purposes a small negative, such as the 35-mm or No. 127, will yield a satisfactory enlargement. Most pictures are printed at one- or two-column width on a three-column page, and good-detail blow-up to full-page width, about 7 inches, is readily available from film materials in these sizes.

The long-time popularity of 35-mm photography for magazine journalism stems not only from the cheapness and quality of film materials, but also from its adaptability to a great variety of exposure conditions.

Large-aperture lenses, which would be prohibitively expensive on larger film cameras, permit hand-held shots in poorly lighted places (down to 4 or 8-foot candles). Fast shutter speeds—0.001 sec is not unusual—stop fast motion where the lighting is adequate.

*Lenses.* Interchangeable lenses of different focal lengths help overcome difficult camera-location situations. For example, it is frequently difficult to get far enough away from a large object to get it all into a normal-angle photo. Use the wide angle. The wide angle also can eliminate undesirable foreground details, like utility poles or nonpertinent equipment, by bringing the camera closer in to get the same photo image. Another wide-angle lens use is to reduce or eliminate converging verticals where the standard-focal-length camera would have to be tilted upward to get all of a high object. With the wide-angle lens, the camera can be held level or nearly so; extraneous imagery in the foreground can be cropped for publication.

Wide-angle lenses also may be used for "interesting" visual effects: sharply receding horizontal lines in vertical planes nearly parallel with the vertical plane through the camera axis; foreground figures that appear gigantic against distant backgrounds.

**Figure 7. Wide-angle lens can gain startling perspectives, emphasizing foreground details in their background context.**

**Figure 8. Long-focus lens achieves a foreshortening effect, bringing together subjects strung out along a distant line.**

An opposite effect is available with long-focus lenses. Nearby and distant objects appear to be closer to each other because their relative sizes appear to be nearly the same. Thus a low-angle photo down a street carrying ordinary traffic can be made to look as if the cars are bumper to bumper.

As an oddity, such a picture might be useful to dramatize a problem in traffic engineering, but for most article-illustrating uses the long-focus lens would merely overcome a camera-position problem: a detail within some operating machinery; a television antenna atop a tall tower. And because the detail sought will also appear (in smaller size) in a normal-lens photo,

**Figure 9. Shallow-focus range at widest aperture permits concentration on desired detail and blurring out of unwanted images at other distances.**

the normal use of the long-focus lens is to get the detail on a larger piece of negative, and hence to enhance its capability for clear presentation.

Another use of the long-focus lens is to de-emphasize unwanted details by deliberately exposing them out of focus and thereby blurring them. This effect is available with all adjustable lenses, but it's most pronounced with long focus. The long-focus lens characteristically has a short depth of focus, even at its smallest aperture. At its widest aperture this lens requires careful focusing to get a sharp image at the intended object's distance. Details both nearer and farther away will be indistinct. This technique, when available, will save the expense of air-brush hazing and avoid the expression of artificiality that air-brushing imparts.

One accessory that is useful when an object about which you write is small is an extension bellows or tube for taking close-ups (closer than about 3 feet). This device positions the lens farther away from the focal (film) plane, beyond the reach of the normal focusing mechanism. Needless to say, precise focusing is essential, and the object must be nearly a plane (a printed electronic circuit would be suitable) unless long exposure time can be used. A single-lens reflex camera is good for focusing; otherwise the extension device must be calibrated.

## Indoor Photography

When you have decided what picture you need, and your equipment is ready, the rest of the job may be easy or difficult, depending on the accessibility of the objects. In your office, laboratory, or other indoor place, if the objects are at hand you have little more to do than arrange them and assure proper lighting. Seek a neutral background, remove extraneous objects, and use shadows to accentuate features. Oblique lighting makes shadows; if any pertinent details are in deep shadows cast by the oblique light, provide lower illumination for those details from a secondary source. Expose for the lower illumination; most film has sufficient exposure latitude to retain the highlights.

Unless you need to stop action, don't use flash—at least from the camera position. Direct flash from the camera position gives photos a flat, garish appearance; object relief casts slight black shadows because the light source is only a few inches away from the lens, but the shadows often are where you'd rather not have them. Primary and secondary flash from oblique angles can be satisfactory, and "bounce flash"—flash illumination of nearby walls—is a technique that gives agreeable results if the photographer plans the effect correctly.

In most cases, however, light from a window, plus an ordinary lamp or pair of lamps can do the job. With a fast film emulsion and a large-aperture lens you can hand-hold your camera. Of course a tripod is a convenience for any still-life photography, and it's a must when shutter speeds below $\frac{1}{30}$ sec are required. (Personally, I'd rather not hand-hold at slower than $\frac{1}{200}$ sec.)

Another consideration in arranging your picture is its expression of the object's size. If the object is likely to be unfamiliar to the reader, or if it's something that is similar throughout a wide range of sizes (crushed stone, woven fabric, an internal combustion engine), something of familiar size should be added to the picture. A lead pencil, a man's hand, or a whole man are common additions. The man can point to an important feature of a machine, or he may appear to be operating it.

## Field Photography

If you must go to the field, presumably outdoors, for your photos, more careful preparation is necessary. In addition to preparing your shooting script you will have to determine that field conditions will be right when you get there, and you will want to be doubly sure that you take along the right equipment.

Let the field people know your plans, including the types of pictures you expect to take, and give them tentative dates. Ask whether construction progress or operational conditions will be right for your photography; it's frustrating to find the plant feature you're interested in shut down the day you arrive. Then watch the weather; hope for a clear day; pray that it won't rain or snow.

If you're going by car, pack all the equipment you can lay your hands on. Don't forget the flash; bright sunshine casts deep shadows that can need secondary illumination. (Not too close, though, or you'll lose the sunlight's natural effect.)

A shoulder bag is a good thing to take along because it keeps items you're not using in one place and saves wear and tear on your pockets. More important, it leaves your hands free for the many manipulations of photography and for climbing to good viewpoints.

If you're using public transportation you'll want to minimize bulk and weight, in that order. You still want all the flexibility you can manage, so you'll be thankful if your equipment is 35-mm-based. First on the list is the camera and a good supply of film; then a light meter. A wide-angle lens is likely to be more useful than a long-focus lens, but take the latter, too, if you have it.

If supplementary view finders are necessary, don't forget them. A lens shade will save you some awkward moments if your timetable goes wrong and you have to shoot toward the sun. Flash can be left behind unless you expect to illuminate deep shadows or you plan to go into a tunnel or a mine. Filters, camel's-hair brush, cable release, and other small items are optional, too, but take them along; they won't get in the way. A tripod, on the other hand, is a nuisance to travel with. Leave it at home unless you're sure you'll need its steady support and you can imagine no substitute.

A second camera is carried into the field by some writers who like to get both black and white and color. Others will take the time for film changing, which is feasible at any place on the roll with most 35-mm cameras. (Don't rewind the film all the way into the cartridge, and do mark down the number of the last frame exposed.)

## Should You Standardize on Color?

Possibly more efficient, even though more costly for film and processing, is all-color-negative photography. As stated previously, the publication can use color prints for black-and-white reproduction—and it may want to reproduce the color. One caution, however: color films have significantly lower emulsion speeds than the faster black and whites, and they have less exposure latitude; hence shadowed objects will be less accessible and lighting will have to be relatively uniform on any view. Fill-in flash procedures must comply with the film maker's instruction about compensation.

## What to Expect in the Field

When you get to the field you'll be subject to site regulations about safety and security. The person in charge probably will hand you a hard hat, and he may detail a man to accompany you. This man can help in a number of ways. He can appear in photos to give them scale, operational reality, or human interest. He can remove extraneous items from view, and he can correct unsafe conditions. Contractors, industrial establishments, and others are sensitive about showing unsafe practices; they're among the first features (along with classified items) that the public relations department looks for when photos are submitted for approval. Examples of unsafe practices are ladders that are not tied in place, scrap lumber with exposed nails, combustible rubbish piles near buildings, people smoking near fuel pumps, and men without hard hats.

When you go into action, get more than one shot of each picture you want. Vary the angles. Look for better lighting and vantage points; ask permission to go up there (but don't ask to ride the crane hook). Assuming good exposures, your photo quality is limited only by the agility of your imagination and your body.

And while you're there in the field, try for some last-minute information on your topic—performance data that haven't been reported back to you yet; solutions to installation problems that will improve the next design; unforeseen benefits that readers ought to know about. These are matters that you would cover during a routine inspection trip in the field; now you can sniff at them with your nose for news.

# 10

# EDITING PRACTICE

"I have no pride of authorship," the contributor tells the editor. This commonplace assertion is supposed to assure the editor that no matter what he does to the submitted material the author won't mind. He doesn't fool the editor, and if the author is experienced he doesn't even fool himself.

Pride of authorship must be a condition if not a motivating factor in all but the most coercive acts of authorship. The disclaimer, then, is a face-saving device to discount the almost certain fact that the editor will tamper with the text.

Not only will he change things like capitalization and abbreviations to make them conform with his magazine's style, he will also be likely to revise sentences and paragraphs to make them fit his ideas of organization and style. Worse still, he may replace the author's pet words with pet words of his own.

This process whereby a manuscript gets altered to some extent is not called tampering, of course; it is called editing. And the editor, as we already have seen, is not a malevolent being to be classed with life's necessary evils. Rather, he is a fair-minded and skilled craftsman in the printed presentation of ideas.

Moreover, he is happiest when the contributed article is so well written and styled that it needs hardly any editing. For one thing, the good one gives him more time to work over the less readily usable one, assuming that their informational values are the same. The good one will be scheduled for publication sooner, too, no matter how impartially the editor is supposed to treat incoming material.

## EDITING APPROACHES VARY

There are extremes of editing practice among magazines, of course.

## Total Rewrite

At one end of the range, certainly not the lower end, but discomfiting to painstaking authors, is the publication that prefers to rewrite everything. This magazine must be blessed with an abundance of editorial manpower (and advertisers); it also must have some reason for total rewrite: achievement of uniform writing style; belief that all contributions should be boiled down; or merely some tenet that total rewrite is best.

## Laissez-Faire

At the other end of the range, certainly the lowest, is the publication that prints, practically as received, any material it can lay its hands on as long as it's pertinent to the field served and reasonably original and up to date. The editor of this publication must be badly overworked; it's unthinkable that he would be lazy or incompetent.

## The Broad Middle Ground

Happily, most editors operate well within the extremes of editing practice. They judge incoming material fairly and present it in the best available fashion.

The main influences on editing practice, relative to contributed articles, are the informational value of the material and the availability of space in which to present it.

## Informational Value Affects "Amount" of Editing

Value means value to the readers and therefore value to the magazine. Contemporary significance within the magazine's field is essential, as is a considered need for coverage of the subject to round out or balance the magazine's over-all presentation. If your article is on a currently overworked subject, however significant the subject is, the editor may want to strip it to its barest essentials: its new facts or conclusions, together with the briefest of statements placing them in contemporary context. On the other hand, if he feels he hasn't published enough in your subject area, he'll give your article prominence, and thus round out his service to the field.

## Space Availability Is Influential, Too

Space considerations influence editing in at least three ways. One influence is the page budget. Most publications allocate a standard number of

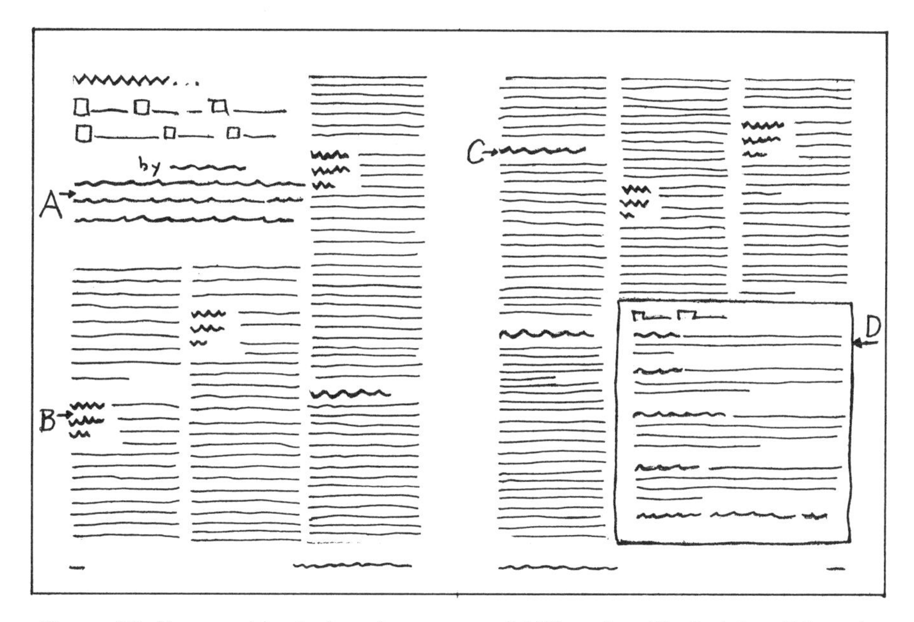


**Figure 10.** Typographic devices improve readability of unillustrated articles. A. A blurb under the headline presents the main idea(s). B. Side head (or C, center head) flags subordinate idea. D. Boxed-in material presents key quotes or tabulates important data.

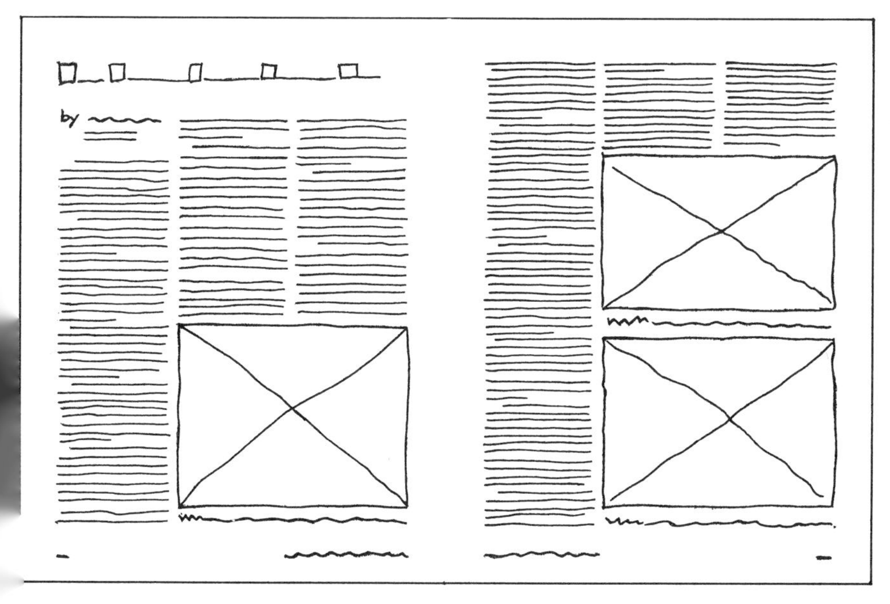


**Figure 11.** "Normal" article layout, including one or two illustrations per page, utilizes short captions that direct attention to the text. If captions, when read consecutively, outline the principal story line, so much the better.

editorial pages in some ratio to normal advertising pages (1:2 to 2:1) and increase the budget for issues that are patronized heavily by advertisers. However, the budget should be fixed sufficiently in advance of each issue for the editor to plan his presentation.

A second space influence on editing practice is related to the overall concept of readership. For maximum readership of given material there is an optimum length. For average technical articles, this length is about two pages in the magazine, or about 2400 word equivalents. If an article is not illustrated, it should be shorter (and it should be dressed up typographically to relieve the visual monotony). Well-illustrated articles can run longer than two or three pages without loss of attention, but the pictures and captions should cover all important points.

Time urgency is the third space influence on editing practice. The article you submit may be so timely that the editor will want to squeeze it into an already planned issue. More likely than not, he will make the room by removing a story that is shorter than yours should be. So he'll boil your article down to fit the space. The skillful editor will treat your article as a trainer would treat a prizefighter making weight—he'll cut the fat and keep the muscle.

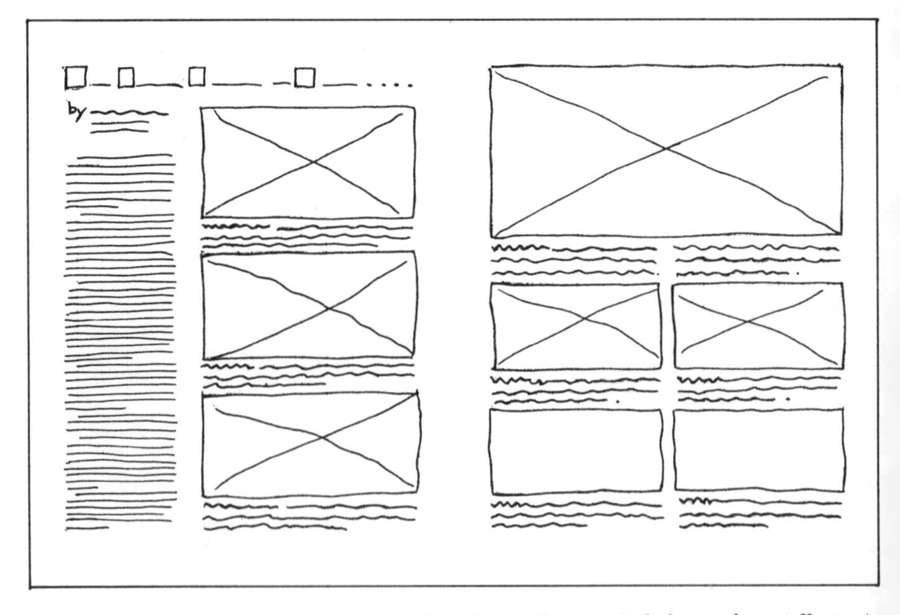


**Figure 12. Simple picture-story layout lets illustrations and their captions tell most of the story. Text summarizes and furnishes fill-in data that are not directly related to the photos.**

## EDITING PROCEDURES

Editing procedures comprise planning and scheduling, writing and editing, and production. Large editorial staffs divide this work, planning and scheduling being a high-level function, writing and editing being an intermediate-level function, and production being a lower-level function. Responsibility remains with the top. The Editor and managing editor handle planning and scheduling; the Editor supervises writing and editing; and the managing editor supervises production.

You, the author, have an interest in the editing process at all levels.

### Planning and Scheduling

From the moment the editor knows about your article it is an element of his planning, assuming he wants the material. The sooner he knows about it, the sooner it can be published, so get in touch with the editor as soon as you see your way clear to writing the article. Tell him what it's about and when you can deliver the text and illustrations. Some agreement on length is desirable, too.

This information about your article is input that goes with page-budget data into a planning program that guides the magazine's scheduling. The program contains a statement of the subject areas covered by the publication, the relative importance of each subject area, and the kinds of presentation utilized. The kinds of presentation may include news, editorial comment, letters to the editors, new products and literature, personals and obituaries, and technical articles. Certain constraints or commitments may be included, such as the requirement of one full page for editorial comment, at least two columns of letters, or a fixed number of news pages.

Thus the planning process is analogous to many problems that are put through computers. The hardware here, however, is the living editorial brain; the software is a combination of written and otherwise known policies, habits of logical thinking, and decision-making techniques.

The output from this "computer" program includes as a major element a schedule of technical articles for future issues.

### Schedule Display

The schedule usually is displayed in some manner near the managing editor's desk (Figure 13). Basically, the schedule displays each article as a unit, grouped with the other articles slated for an issue. Tabular form on

**Figure 13. This managing editor's schedule board has corner-recessed studs that hold article tickets the way "art corners" hold photos in an album. Vertical columns are issue months. Each horizontal line presents one editor's scheduled tasks. "X" on ticket means the completed article has reached the managing editor. (Courtesy of Plant Operation and Management.)**

paper, with one column for listing the articles in each issue, will do, but greater flexibility usually is sought through use of some form of scheduling board. A typical board has several columns of transparent pockets the size of theater tickets. The top pocket in each column displays an issue date, beginning at the left with the next issue and proceeding to the right with the dates of successive issues. Catchwords or phrases are written on blank "tickets" to identify each article and indicate its length. These tickets are inserted in separate pockets below the dates for which their articles are scheduled.

A refinement of the system is to subdivide the board into horizontal bands representing individual editors' tasks or the principal coverage areas (subfields) of the publication. Thus balance of coverage can be seen at a glance. Lopsided coverage can be corrected by switching articles from issue to issue; in other words, by rescheduling.

Other reasons for rescheduling would include pre-emption (by unexpected, important material) and late arrival of expected material.

## Writing and Editing

When your article arrives, assuming it already has been scheduled, it enters the editing phase of the editorial process outlined above. This phase is called the writing and editing phase because there is a broad overlap of writing and editing. Moreover, most of the editors who would edit your article also report, write, and edit articles of their own. (Called "staff-written," these articles may be either by-lined or unsigned; practices vary widely among magazines.)

## General Appraisal

Your article is read first by a high-level staff member to confirm its usefulness to the readers and its conformance with his original understanding of your concept. He also may jot down ideas for subsequent treatment; changes in emphasis; need for extra data; special illustration techniques; layout.

If extra information is needed, the editor should request it from you. In his letter to you he could be expected to discuss the treatment ideas he has in mind, and he might enclose your manuscript so that you can work it over yourself. He won't return any of your illustrations unless he wants you to clarify your point about one or more of them.

## Detailed Edit

If the editor is satisfied that he understands your message and if he feels that it is substantially complete, he will proceed with a "desk edit." That is, he will mark the manuscript wherever necessary to make it conform with his magazine's style in spelling, punctuation, capitalization, word compounding, abbreviation, and numeralization; he'll mark straight lines under words to be italicized and wavy lines under words to be boldfaced. He may also change or insert a word here and there, transpose a few phrases, or delete a redundancy or two if he thinks these measures will make your meanings more evident.

## Rewrite the Opening?

Somewhere in this process the editor will take a hard look at your first few paragraphs, seeking ways to improve their appeal to readership. If he decides he can improve on your opening, he'll make any changes he sees fit; rewrite is not uncommon for the opening, but here as elsewhere throughout the manuscript most editors prefer to keep intact as much of your writing as possible. They not only respect your authorship; they know that needless rewrite consumes valuable time—and affords unnecessary opportunities for originating mistakes and developing changed meanings.

## Preliminary Layout

Now for a look at the illustrations. The editor will decide which illustrations are essential to your story and how much space they ought to occupy for optimum understanding or eye appeal.

Desirable illustration space is added to text length to yield an estimate of the number of published pages the article would require at this stage of the editing process. The unit of measurement is a column inch. Illustrations usually are one, two, or three columns wide, so the number of columns is multiplied by the depth corresponding to the selected width to get the number of column inches for each illustration.

Double-space typescript sheets normally run about 25 lines of 60 characters each, or 300 words, using the standard 5 characters (including spaces) per word. A typical technical or business magazine runs about 1200 words on a three-column page. If the columns are 9½ inches long, there are 28½ column inches on a page and your manuscript will run 42 words to the column inch, or about 7 column inches to a sheet of copy.

If the article turns out to be longer than the number of pages initially scheduled for it, the editor can either revise the schedule to get more space or shorten the article. Shortening the article may involve deleting or reducing the size of illustrations, or rewriting text, or both.

If the article won't fill the allocated space, illustrations may be added if they're available, or a "filler" item may be inserted on the last page. It is unusual to expand the text to fill more than a small number of lines, because unnecessary text impairs readership.

On completing the length estimate and finalizing the space allocation, the editor is likely to sketch a layout for each page. Within a rectangle of approximate page proportion but not necessarily the page size, he asserts his preference for a left or right-hand page start, blocks out illustration locations, and indicates the headline treatment.

## Headlines

Some publications have a small range of standard headlines for articles of various length. An example of such a schedule is on the facing page.

Other publications' use of display type is limited only by the ingenuity and good taste of the editorial staff.

| **Article length** | **Headline Styles** |
|---|---|
| Three or more pages | 36 pt C&lc, one line on 3 cols.[1] |
| Two pages | 24 pt C&lc, two lines on 2 cols. |
| One page | 24 pt C&lc, one line on 2 cols., or 18 pt C&lc, two lines on 2 cols. |

Whatever the headline treatment, this stage of the editing process is an appropriate time to write the headline. It may be short, to arouse curiosity, or it may be long, to tell the busy reader at a glance whether he need read further.

An example of the short head is "Boston Unbound!" written for a story about long-range traffic and transportation improvement plans for the Hub of the Universe. A tell-all long head might be:

**"Steam-Dumping By-pass Chamber
Designed for Once-Through Boiler"**

The shorter headline frequently is followed by a boldface sentence or two about the story's highlights. Journalists will tend to call this bit a blurb, a word they also apply to some other types of brief editor-written material (short announcements by the publication, statements about upcoming articles, etc.).

If the publication uses side heads or center heads within the text, they're written into the copy at this time.

## Caption Treatment

Captions, too, get the editor's detailed attention—at last. Looking at the layout sketch, he notes any situations where caption length is important from the visual standpoint. For example, captions under adjacent illustrations should have the same number of lines; the longer of the two probably will be cut to the length of the shorter, assuming that all data essential to understanding can be preserved. (Remember, you wrote the caption fully.)

The editor will consider how captions, illustrations, and text work together to tell a consistent and understandable story, and if he needs to shorten the story, he will decide how much of the story-telling burden can be shifted to illustrations and captions.

## Illustration Processing

Detailed work on illustrations can begin at this point. Photos will be cropped and scaled for exact size; any "doctoring" that's needed will be

[1] 36 point (36/72 inch or 1/2 inch high), capitals and lower case, one line of words extending across three columns.

specified for the artist; instructions will be prepared for drafting drawings, graphs, and charts—all as described in Chapter 9, on illustrations.

### Manuscript Typing

After a final read-through of the manuscript the editor will get it typed, with at least one carbon copy. For typing work, many magazines have special forms printed in an inconspicuous tone on 8½ × 11 inch yellow or lightweight bond paper. A heading area at the top of each sheet bears the magazine's name and spaces for the author's name, a short title, the editor's initials, the page numbers, and one or more dates (date typed, date to printer, for example). Double-space line positions are numbered at the left from top to bottom of the typing area, and vertical guide lines mark left- and right-hand margins. These margins are spaced so that one line of typing equals either one or two lines of type set in the magazine's regular text font, to afford close estimating of the article's length.

The rules about starting the first page part way down and typing captions and tables on separate sheets, cited in Chapter 8, on text preparation, are observed.

### Carbon Copy to You

When this manuscript comes back from the typist, the editor should send you the carbon copy to review for factual accuracy. If he has changed your writing substantially he may feel compelled to discuss the changes. He'll probably tell you when the article is scheduled and give you a deadline for returning the carbon copy. This story-checking step may be omitted if the editor has decided to run your story substantially as written. If he omits the step for any other reason, such as haste to publish, send your next story to another publication.

### Copy Marking

When your copy comes back, the editor makes your corrections on his original, then marks type fonts and widths on headings, text, captions, and tables. A font comprises the type of a given face in one of its sizes, face being a name, such as Bodoni Book, and size being a number of points (1 point = 1/72 inch), such as 9 point. Slug thickness also is specified if the line spacing is to be more than the type size. Width is specified in another unit that is peculiar to printing, the pica (= ⅙ inch). Thus, 9 on 10 point Bodoni Book by 13 picas would specify 9 point type on a 10 point spacing (7.2 lines to the inch), set 2⅙ inches wide. An illustration of this practice appears in Appendix B.

At about this time the editor will have all illustrations in "camera-ready" form. If his printing process is letterpress, he'll finish marking the illustrations for the engraver and order the cuts: half-tones for photos and other toned artwork; line cuts for line illustrations. If the process is offset, he'll do a similar marking job for photo-offset plate making.

## Production Has Begun

Actually, somewhere in the two preceding paragraphs, the over-all procedure moves from the editing phase to the production phase. It is an act of production to send the marked manuscript to the typographer; it is an act of production to send the illustrations to the engraver or the negative maker.

The typographer sends back "galley proofs" of everything he sets. The proofs are read and returned to the typographer for correcting; then he sends back either dummy proofs (for letterpress, usually on colored paper) or reproduction proofs (for offset, a sharp impression on high-quality white paper).

## For Letterpress, a Dummy

The dummy is the letterpress compositor's instruction sheet for magazine makeup (Figure 14). Type and illustration proofs are pasted together in the positions they will occupy on the magazine's published page. Paste-down can be done on a back-issue page, which provides guidelines for column position and length, but most magazines have special dummy forms printed in a light tone on sheets that are larger than magazine-page size.

A good dummy form shows column outlines and page trim lines. It has the magazine's name in its normal printed position and it provides spaces in the proper locations for page number and issue date. The form may also have a dashed line ⅛ inch outside the trim line to indicate where the edges of bleed cut proofs should be pasted.

During dummy paste-down it usually happens that the text is too long or too short. The editorial production worker's first reaction is to wish that there was such a thing as rubber type. Then he cuts or adds, obtaining competent assistance if necessary. For cutting he may trim a photographic illustration if it won't impair the value of illustration or the overall visual effect of his layout. The production man, it should be noted, is quite likely to have strong opinions about the esthetics of graphic presentation, the role it plays in reader psychology, and precise ways to achieve maximum readership.

The rest of the letterpress procedure comprises the following steps: typesetting for space-adjustment alterations; makeup by the compositor (placing type and cuts in positions specified by the dummy); reading page

**Figure 14. Editorial worker prepares instructions for letterpress page make-up, by pasting type and illustration proofs in desired positions on dummy sheet. (Courtesy of Water & Wastes Engineering.)**

proofs; typesetting for page proof corrections; final makeup, lock-up (clamping the cuts and heavy type metal in a steel frame) and make-ready (fine adjustments of type height to obtain a uniform impression); (stereotyping—making a cylindrical mold of the type—if a rotary press is used); and the press run.

## For Offset, a Mechanical

For the offset printing process, reproduction proofs of type matter are affixed to heavy cardboard (Figure 15). Here position is even more important than it is for letterpress, because the paste-down position *is* the position in which each image will appear on the final printed page. Illustration positions are delineated at size and proportion scaled from the originals, and identifying symbols are marked in the spaces. Corrections and adjustments to type can be pasted over the initial paste-down. The products of this step—reproduction proofs on board—are called "mechanicals." Originals of illustrations accompany the mechanical unless negatives of the illustrations have been made already.

The rest of the offset procedure comprises the following steps: making a same-size photographic negative of the mechanical; (making and) stripping in illustration negatives in their designated positions; making a blue print (or brown print) from the negative for page-proof reading; typesetting for corrections; photographing and stripping in corrections; platemaking ("burning" the plate); and the press run.

**Figure 15. Illustration department member prepares camera-ready mechanical for offset printing, by affixing reproduction proofs of type and line art work in exact position on clean, stiff board. (Courtesy of Water & Wastes Engineering.)**

After that, there's folding, trimming, binding, mailing—and the reason for the whole process: educating, entertaining, informing, or impressing the reader.

After publication, the editor may have one or two more duties relative to your article. If the magazine accords an honorarium he'll see to it that the check is issued to you (many publications pay sooner); and he'll return illustration materials to you if you asked to have them back.

Your own reaction when you finally see your work in print will be primarily one of pleasure. You may even be willing to admit that you do, indeed, have pride of authorship.

# 11

# HELP IN YOUR OWN ORGANIZATION

How often have you thought, "I wish I had the time to do an article about this" ? And have your thoughts been forced into the open by the question, "Why don't you write about that? "

Not enough time: is that it?

## HELP FROM PUBLIC RELATIONS

Run, don't walk, to the public relations department. If your organization doesn't have a formal public relations program, go to management and ask for help. You'll get more than sympathy, because the organization stands to gain as much as you do, and possibly a lot more. Your writing is good publicity, a beneficial element in image building.

Help in your own organization may be as simple as typing and drafting services; perhaps that's all you need. Or else you may need—and get—complete ghostwriting help.

### Ghostwriting Can Be Justified

You need not recoil from the term "ghostwriting," any more than you should lower your voice when you say "publicity." Admittedly, publicity has a popular connotation of press agentry and showy attention getting. But publicity also functions usefully—particularly in technical and business fields, where the needs exists to disseminate widely all kinds of information that will improve civilization. Publicity, or rather the obligation to publicize and the benefits therefrom, is what this book is about.

As to ghostwriting, it frequently happens that the man who has something worth writing about also has other things to do that are urgent. Competent writing help frees him to do the urgent, which presumably makes him and/or the organization more money than the ghostwriting costs.

## Public Relations Functions

The public relations function of an organization has at least three facets: a planned-image program, control of employee contacts with the public, and troubleshooting.

Troubleshooting is unlikely to concern the writer of articles, unless he unwittingly discloses information that is confidential or otherwise against the organization's best interest. Moreover, control of external contacts—usually a blanket requirement that public utterances be cleared by management—is supposed to preclude the need for troubleshooting.

The planned-image program subdivides into paid advertising, community services, speeches, and publicity. Publicity breaks down further into publicity initiated by the organization (planned) and that which is initiated by publications (unplanned by the organization, but usually stimulated by organization-initiated publicity).

## In-House PR versus Agency PR

Instrumentalities of public relations may be mostly in-house, substantially outside, or in some intermediate proportion between the two locations.

Many large organizations perform all first-hand functions but ad placement. Bona fide advertising agencies get a fee from magazines, traditionally 15% of space rates, for placement of ads, so there is little incentive for a company to perform this function.

Another normal exception to all-in-house public relations functioning is art work. Independent "studios," usually one or two people, abound in most cities to prepare illustrations and make layouts and mechanicals. Advertising agencies use the independents, too, keeping their own art staffs at a size to handle the base work load.

In publicity, a typical example of part-inside, part-outside operations is the procedure wherein a company employee, who may have the title of public relations manager, writes news releases and turns them over to a public relations agency for duplication and mailing. The mailing list may be developed by the agency from a list of subject categories supplied by the company.

The corresponding example, for the substantially outside operation, goes something like this. An official of the company, possibly the sales vice president, calls his contact in the agency and asks him to come over and get a story. The agency man listens to the release idea, canvasses available background data, interviews appropriate people, and writes the release. After its approval he carries through the reproduction and mailing.

The point is, there is a range of available means of accomplishing a public relations task. The choice depends on the economics of an organization's public relations needs and the preferences of management. Economic factors are based on the kind and quantity of public relations effort that management deems necessary to accomplish organizational objectives. Presumably these efforts can be translated into schedules of man-day needs. A steady flow of effort would favor in-house capability; marked fluctuations would indicate heavy reliance on outside help under some sort of fee arrangement.

## How PR Helps You

Regardless of the instrumentality of organizational public relations, the people who perform these functions can help you at any or every step of the article-writing process.

Start with the idea. Public relations will help you appraise the publication possibilities of your idea. You go to the public relations man and tell him what you have in mind. You describe the idea, tell him what you know about its significance in its field—the background of prior developments and contemporary efforts by others in the same line. You also tell him what you believe will be the advantages to be derived by the company from publication of the idea. If you have a publication in mind, you name it.

The public relations man will consider first the benefits to the company. (Bear in mind the identity of your interest and the company's interest.) He'll corroborate your thoughts about these benefits, and he may come up with others. He may suggest emphasis of certain points, and he also may advise caution about some other aspects of your idea. Caution at this preliminary stage is likely to pertain to timing. Premature disclosure of innovations can impair expected competitive advantages. Benefits claimed for untested developments can become invalid.

Your thoughts about the appropriate publication may be modified, too. Public relations men are accustomed to evaluating the respective merits of whole groups of magazines. They also like to place different versions of articles in two or more publications that don't overlap each others' readership. However, you're the author. You can insist on trying a certain magazine.

## Editors Respect PR Men

You also can approach that magazine's editors; or public relations may do it. Your public relations people and the editors of the principal magazines serving your field know each other. It's their business to know each other. The chances are good, too, that they respect each other. The tyro editorial worker may have a schooled-in suspicion of PR flak, and he may be justified if he's in general publications, where the rewards of consumer product promotion may warrant ballyhoo, but he'll soon find some of his best friends among the hard-working, no-nonsense public relations people in organizations like yours.

It is the public relations man's business to cooperate with the press. He knows his own organization, its people, and its lines of authority. When an editor wants a story he approaches public relations first, explaining what he has in mind and possibly suggesting which people can help. Public relations arranges interviews, assembles data, even supplies the story if that is the way the editor wants to handle it. The net effect is to save the editor the time and trouble of making numerous contacts throughout the organization—and probably ending up in the public relations department anyway.

## PR Will Do Your Selling

This go-between function of the public relations people is a two-way street. They'll be happy to place your article: present your idea to the editor, gain his tentative approval, and bring his suggestions back to you. In short, public relations will do your selling job for you, saving you time and possibly doing a better selling job than you could. This assertion is made on the assumption that the editor is not already a friend of yours. It would be wasted motion, or possibly slavish adherence to "policy," to interject public relations between you and your friendly editor at this stage.

## PR Will Do Your Writing

There would be nothing wrong about introducing your public relations people to the editor in the next stage, however. Writing is the stage of article production in which public relations can save you really significant time. Not only will you liberate large chunks of actual writing time, you will also shorten drastically the time elapsed between idea crystallization and draft completion.

For busy men, particularly those whose workday consists of constant interruptions, it is difficult to assemble consecutive minutes to concentrate on any knotty problem. Writing requires "tooling-up" time—review of the outline and what was written previously—and writing time. For this writer, any period less than a half hour is useless. An uninterrupted period of 45 minutes is a satisfactory minimum block of time.

During working hours pressing "projects" imposed by fellow workers have a way of pre-empting time that would be suitable for writing. It then becomes necessary to find the time during evenings and weekends. Boy Scouts, professional societies, recreation, and home maintenance take their toll from this so-called leisure time. Besides, writing is hard work. The result is that the determined writer is lucky to get three or even two pen-and-paper sessions in a week. (That is enough to produce a two- or three-page article in 2 or 3 weeks.)

The less-than-determined writer is out of luck—unless he has public relations in his corner rooting for him, demanding action, and even doing his writing.

## A Modus Operandi for Writing

A typical full-writing service for a contributor to the literature works in the following manner, assuming the existence of an outline and an expression of interest by a publication.

Public relations sits with the author and takes notes as the author talks about the outline. Automatically, one problem of gaining writing time is solved. The public relations man *is* the interruption. Telephones can intrude on this time, of course, but the visitor will hold the floor.

As he talks, the author feeds in illustrative and background materials. Hopefully, he has assembled them for the interview, but by merely naming them he brings them into the open, so to speak, because the interviewer will see to it that the author or someone he names produces the missing materials as soon as possible.

The interviewer asks questions to clarify concepts and bring out points he suspects will be significant. He can be expected to be familiar with either the author's kind of work or his organization's activities, or both, so it should be unnecessary to use extra time for "education." This happy likelihood is attributable to the state-of-the-art of public relations. The demand for ghostwriting talents is so great that all kinds of specialists work in the field. Their collective experitise is analogous to that of the technical and business magazines' editorial staffs. In fact, a typical course of the technical journalist's career is school (to industry) to magazine staff to public relations agency (or department).

When the public relations man finishes the interview, he returns to his typewriter. In a few days he delivers a double-spaced draft to the author.

## Pride of Authorship Blossoms

Now an interesting transformation takes place, assuming the draft was prepared with reasonable competence. The author hops right into the job of reading and correcting. Time is no longer a problem; he finds time or makes time. There's something about the presence of words on paper that instantly brings out the editor in authors who shy away from the blank sheet. Moreover, pride of authorship develops quickly at this point.

This pride of authorship is proper. The ideas belong to the author and he uttered many of the words in the interview. Moreover, the writing style may very well be his. Many a good ghost has, in addition to his ability to understand subject matter, a remarkable knack for picking up another person's style.

## PR Men Know Illustrations, Too

Less remarkable, but nevertheless highly useful, is the public relations man's proficiency with illustrations. He will advise on the adequacy of materials in hand, suggest improvements, take steps to obtain better illustrations, and plan sophisticated graphic presentations. Later he may even collaborate with the editor on the production of art work.

## Handling Clearances

But before the illustrations and manuscript get to the editor, clearances may be necessary. In-company, the public relations department would handle clearances regardless of how the article was produced. Normally public relations screens articles to assure that they do not contain anything inimical to image or security. A further requirement frequently is that a company officer see all material proposed for publication. Other clearances —for use of information developed for clients, for example—require correspondence that public relations can handle to save still more time for the busy author.

## Help on Revisions

After submission of the article, public relations can continue to serve by helping the author with any revisions requested by the editor, such as artwork cooperation, mentioned above, and by keeping informed on the article's scheduling.

It is reiterated at this point that most of the public relations services described above are optional, being substitutes for do-it-yourself author tasks. Furthermore, any or all of the services may be utilized to complement the author's work in the manner that is most suitable and convenient to him.

## HELP FROM YOUR PEERS

This discussion of help in your own organization, although it is largely and properly a discussion of assistance from the staff functional area called public relations, would be incomplete without mention of help that line people can render.

### The Fellow Employee as Critic

The most obvious helping role is that of the critic. The fellow worker who is familiar with your subject matter can help all along the way: he should be a good sounding board for preliminary discussion of the article ideas; he can review your outline for completeness of coverage on important points; he can suggest choices of illustrations; and he might help by reading your draft. The last step will help more in the area of completeness and correctness. Someone less familiar with the subject should judge the material's understandability. This second person, also in your organization, should be a "typical reader."

Presumably the typical reader has the technical background for understanding the text; if he doesn't understand it his failure must be attributed to deficient writing. His technique is to read at normal speed, making a check mark in the margin beside any lines, sentences, or paragraphs that he does not understand.

Actually, "understand" in the previous sentence should read "understand readily." Wherever the reader stumbles—has to reread, struggle for word meanings, and perhaps outguess defective syntax—he should make a check mark.

After he has finished, the reader will perform a further useful service by considering his check marks one by one and writing pertinent remarks: questions that need answering; words that might improve meanings; sentence-structure revisions.

### All Authors Need Lucidity Review

The need for a "lucidity review" is best understood and appreciated by the experienced writer, because he knows well how difficult it is to maintain a detached viewpoint while writing. Whether the subject be his own work or someone else's, he must understand the subject before he writes about it. Becoming thus immersed in the subject, he unconsciously tends to erect little bridges in his thinking about the process or chronology. These bridges become mistaken assumptions of reader knowledge, and thus, paradoxically, they become gaps in the account as it is written down. The reader stumbles over the gap if it's small enough; he comes to a complete, baffled halt if the gap is too wide to jump.

So don't omit the lucidity review—and don't feel badly about the deficiencies it reveals. They're just as normal as cogent writing is. Your reaction should be, "Why, of course," and, because the deficiencies are readily retrievable bits of information, you can insert them without delay.

## Ignore Overconstructive Criticism

Some further words about fellow workers' criticisms. They can be constructive beyond the needs of a lucidity review by suggesting amplifications, changes of emphasis, or new topics. Such suggestions should be considered out of scope at this stage of the game. Hew to the line. Go ahead and publish, and let the chips fall where they may.

# 12

# AFTER PUBLICATION, WHAT?

In a way, publication of an article is like a birth. The event of publication is just about the only finite, predictable result of all the painstaking planning, outlining, writing, illustration, editing—and waiting. After publication, many influences, both hereditary and environmental, determine the fate of the article. That's the way it is with new babies. It's all right to assert good-naturedly, as does the song about F. D. R. Jones, "He'll be famous, as famous as he can be," but neither fame nor any other form of success can be guaranteed.

One thing is certain, however. If the baby (or the article) were never born (published) there would be no chance at all for success. So let's publish.

## LETTERS TO THE EDITOR

One of the immediate types of reaction to a published article is the letter to the editor. The number and nature of the letters about your article are a sort of gauge of its significance, although some topics just naturally stimulate more comment than others.

In their letters readers pick flaws, take issue, amplify information, use your article as a point of departure for plugging their own ideas, or say, "Me, too." Not very many of the flaw-pickers' letters are published; not many of the flaws are important, and many of the alleged flaws are not flaws at all.

A memorable and avowed flaw-picker was Old Fancher, a Linotype operator I knew in my youth. Old Fancher claimed to be one of the original Mergenthaler trainees; he was an incontestable authority on orthography and style, to the continuous embarrassment (and gratitude) of generations of student journalists, and he was a veritable almanac when it came to facts.

At home Mr. Fancher kept a stack of penny postcards by his radio. Whenever he heard what he considered defective grammar or fact he would scribble a correction and address it to the radio station. He didn't talk about results, but it's reasonable to assume there were a few—just as a few flaw-picking letters to the editor get published.

Most other types of letters are appreciated by editors, and it's likely that they will be published at the earliest opportunity. If your article appears to be generating substantial comment, the editor may hold the letters for publication together in one issue. He may solicit your comment on some of the letters, particularly if amplification or rebuttal seems necessary, and publish your new material along with the letters.

A group of letters on one article or contemporary topic may be scheduled for a certain issue, but the normal treatment of letters is as "fillers," which are published when space becomes available. The "Letters to the Editor" section thus may occupy one column or it may run for many pages, usually sandwiched with advertising. Special issues attract extra advertising, increasing the editorial space budget and drawing down the editors' reservoir of fillers. (Other stock filler subjects are new products, manufacturers' literature, personals, obituaries, helpful hints, shop notes—any short item of interest to the audience, provided the item doesn't belong in the spot-news category.)

## REPRINTS ARE USEFUL

The other prevalent postpublication activity is reprinting. A reprint is a device for practically assuring that an article will be read, or at least "noted" by selected people. It removes the article from the competition of other articles and four-color advertisements, and places it face up in the middle of the recipient's desk. This person is supposed to be impressed; if he's interested at all he probably *will* be impressed, because he's likely to have confidence in the editor's (any editor's) opinion of what is worth publishing.

The reprint, therefore, gives an article mileage where it counts—with prospective clients, customers, and legislative bodies, for example. Furthermore, it is with the reprint that the author's employer can reap the most benefits.

### Reprint Production

Most magazines will produce reprints as an accommodation to authors and others, but reproduction facilities are so widely available that it can be quicker and almost as convenient to do the job locally. All that's needed is the magazine's permission and a couple of sets of tear sheets. The magazine will be glad to send the tear sheets along with the permission, which usually specifies the manner of presenting the credit line.

An example of a credit line is "Reprinted from *Power Engineering,* November 1968." Some publications specify inclusion of "By permission of," and the word "Copyright" or its symbol. The name of the publishing company and the expression "All rights reserved" may be required. Usually the credit line goes at the bottom of the first reprinted page, unless a special title page is added, in which case the credit line goes there.

A normal title page has four elements: the title, the by-line, the credit

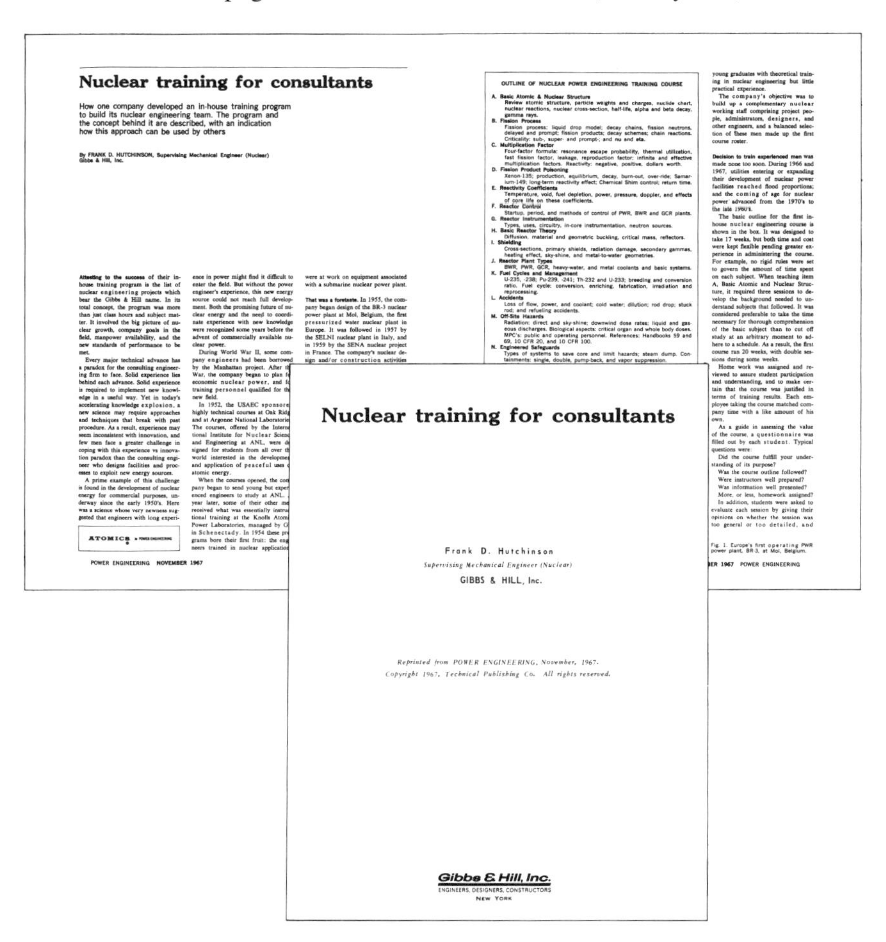

**Nuclear training for consultants**

How one company developed an in-house training program to build its nuclear engineering team. The program and the concept behind it are described, with an indication how this approach can be used by others

By FRANK D. HUTCHINSON, Supervising Mechanical Engineer (Nuclear)
Gibbs & Hill, Inc.

OUTLINE OF NUCLEAR POWER ENGINEERING TRAINING COURSE

A. Basic Atomic & Nuclear Structure
B. Fission Process
C. Multiplication Factor
D. Fission Product Poisoning
E. Reactivity Coefficients
F. Reactor Control
G. Reactor Instrumentation
H. Basic Reactor Theory
I. Shielding
J. Reactor Plant Types
K. Fuel Cycles and Management
L. Accidents
M. Off-Site Hazards
N. Engineered Safeguards

POWER ENGINEERING NOVEMBER 1967

**Nuclear training for consultants**

Frank D. Hutchinson
*Supervising Mechanical Engineer (Nuclear)*
GIBBS & HILL, Inc.

*Reprinted from POWER ENGINEERING, November, 1967.*
*Copyright 1967, Technical Publishing Co. All rights reserved.*

Gibbs & Hill, Inc.
ENGINEERS, DESIGNERS, CONSTRUCTORS
NEW YORK

**Figure 16. Title page dresses up reprint and permits presentation of magazine's double-page spread layout for first two pages of article.**

line, and the logotype of the sponsoring organization. Conventionally these elements are arranged from top to bottom, but other logical arrangements can be attractive, and a second color, either solid or toned, may be applied to good effect.

Preparation of the title page for the offset printer's camera is simple.

*Bernard Gillet (France), C. S. Hsu (Taiwan), and Suphi Cicekli (Turkey) meet with G&H chief designer, John E. Edwards.*

# Why We Seek Foreign-Born Employees

ROBERT H. DODDS, and JOHN EDWARDS, Gibbs & Hill, Inc.

Most technically oriented firms, especially consulting engineering organizations, are finding themselves caught in a manpower gap – a shortage of professionally qualified designers and draftsmen. Unless steps are taken to remedy the situation, the results can be expressed in six words: slipping schedules, decreased efficiency, higher costs. And the consequence of these can be expressed in two more words: lower profits.

**Foreign-Born Workers One Answer**

There are ways of getting around this shortage. Most are expensive in terms of time, money, or both, but we at Gibbs & Hill have found one approach that preserves schedules and profits. That is the hiring of foreign-born workers – primarily, those already in the U. S. who approach us for employment. Our experience has been highly encouraging.

The first point to be made is that a company cannot just sit tight and hope that the shortage will go away because the demand for engineering and design services shows no sign of abating. This demand, in turn, requires more engineers and technicians, both beginners and people whose experience is adaptable to our needs.

Traditional sources of beginners – engineering colleges and technical institutes – spread their gradually increasing numbers of graduates thinner and thinner over ever-broadening fields of opportunity. For example, many students who formerly would have come directly into engineering offices as designers or draftsmen are instead going on into graduate work, much of it in well-financed research. Let's face it, a research project can look much more glamorous to a young graduating student than an eight-hour job with the rather bland title of junior engineer or draftsman.

This is helped along by the professors, who come ever decreasingly from the ranks of practicing professionals, but rather from the bachelor-research-PhD process. They are products of the educational system alone, and they tend to draw more of the same into the system.

Other graduates, upon receiving their bachelor's degrees, go into business administration on the premise that an administrator with an engineering background is a valuable asset. This premise probably is valid for

**Figure 17. Right-hand-start articles may be reprinted without title pages, as above. Credit line appears in margin.**

Paste the article's own title and by-line type in the desired positions; type the credit line in one of the contemporary faces, such as IBM Executive; paste it in place; and cut out the appropriate legend from the organization's letterhead to complete the job.

The pages of the article can be copied without further treatment, or they may be affixed to back-up paper or board. Page numbers can be obliterated by pasting over pieces of plain paper; the magazine's name and issue date usually are left intact.

The decision to print a title page is influenced partly by the article's arrangement in the magazine. A two-page-spread start is a natural for title-page treatment because it is desirable to preserve the original two-page layout effect, making those pages Nos. 2 and 3 of the reprint and leaving No. 1 for the made-up title sheet (Figure 16). A third article page thus would be page 4 of the reprint, produced as a folded sheet of 11 × 17 inch paper. A four-page article with a double-spread start should be reprinted on six pages with title page front and page 6 blank. With a right-hand start, the four-page article would be printed on four pages without title page (Figure 17), or on six pages with a title page whose inside back (page 2) is blank.

Where an article runs beyond its full pages to share space with other editorial material or advertising, the article should be clipped and pasted together on a new layout. Full page width should be preserved. Columns should be of equal length, beginning at the top of the page, so that unused space will be in the lower portion of the page.

That's about all there is to preparing reprints for a printer. All but the smallest commercial offset shops can handle an 11 × 17 inch sheet, and they can print anything their cameras can see. (It should be noted that small-job reprinting directly from the magazine's original type or plates is rare; even when the magazine staff handles the job, separate photo-offset procedure is followed. Volume reprinting of special editorial features, planned in advance of publication, may utilize the original medium, however.)

For companies and authors alike, the reprint adds impetus to the enthusiasm for writing that is generated by the event of publication. Salesmen like to give reprints to prospects and customers; authors receive requests for single or multiple copies, the larger quantites frequently being requested by professors for classroom distribution.

## AFTER PUBLICATION, WRITE ANOTHER ARTICLE

For the salesman a reprint of one article is but one contact, so he likes it when the technical people keep the articles (and reprints) coming. The author knows that if his writing is to bring him sustained recognition he must continue to contribute to the literature. Thus "write another article," may well be the best all-round answer to the question, "After publication, what?"

Fortunately, the second article is easier to write, assuming that the author has something to write about. There are reasons why writing is easier at this point. In the first place, the author has been through the writing-editing-publishing procedure at least once, so he has a first-hand basis for confidence in his ability to produce. Then he has new incentives to write: the pleasure of seeing his material in print; the tickle of an honorarium or an "editorial award"—typically a United States Savings Bond—by his employer. Employers who conduct editorial awards programs have observed that the awards seem to stimulate second articles more than they generate maiden efforts.

Having something to write about need not be a problem, either. The author who felt that he had described all of his work that was worth writing about now may want to amplify or rebut criticism that he stimulated. Or he may discover that another publication would be happy to accept a different version of his story. The second publication should not be a competitor of the first, however, because the principle of exclusive release deserves conscientious compliance. Being a noncompetitor, the second publication would have a different audience; therefore a different set of benefits would appeal to the second audience.

To illustrate, suppose you are the project engineer in charge of the design of a nuclear power plant. Because you wanted to impress electric utility companies most, you wrote your first article for *Electrical World* or *Electric Light & Power.* You stressed benefits of plant-type choice (fuel savings, air pollution control), safety (advanced, simplified "fail-safe" systems), and reliability (redundancy, protection, and interconnection), and also mentioned reactor-containment cost improvement attributable to a new prestressed concrete design.

You barely mentioned the plant feature that will be of greatest interest to a civil engineering audience: the advanced-design reactor-containment structure. For the second audience, the economies of eliminating a steel vessel will be the benefit worth stressing. How prestressing is applied and how it assures vapor-tightness will be the first questions asked by this audience. Matters of plant-type choice, safety, and operational reliability are peripheral (but don't omit them; civil engineers, too, need to know the context of their activities and decisions).

So you have a second article to write about the design. Later on there will be more opportunities: construction stories for one or more audiences; an account of operating experience confirming design expectations.

And it isn't just the project or process type of story that presents opportunities for further writing. "Stories worth repeating" should not be ignored, and state-of-the-art articles can be a fruitful type of activity. Particularly in fast-changing fields, the need for updating recurs constantly.

## BECOME A CORRESPONDENT

Forms of editorial service other than article writing can result from newfound acquaintance with magazines. Editors appreciate brief items about minor technical developments in your field: helpful tools or devices for shop practice; field shortcuts; inspection methods; nomograms. Many publications accept news items that they don't otherwise obtain through staff or correspondent activity: significant technical disclosures made at minor conferences; accounts of disaster damage to engineered works.

The editors will listen to you because by writing for them you have established yourself as an expert in a facet of the field they cover. You may offer to cover spot news within your field and your reach, or the editors may ask you to do it. If a bridge fails in a remote area of the United States the editor of *Engineering News-Record* may ask a local structural engineer who has written for *ENR* to report on the occurrence. (Moreover, the editor himself might see fit to join the local expert at the scene. Important new information frequently is developed from failures, but bringing it to light can require high reportorial skill and persistence. Fear of prestige loss and the specter of lawsuits tends to seal lips and to conceal useful physical evidence.)

Another aspect of the editor's willingness to listen to you lies in the realm of opinion. Your writing in this area can range all the way from a brief letter to the editor, on a small point requiring few facts for back-up, to a full-fledged analysis, written in the reasoned and dignified tones of the editorial page.

### An Editorial-Writing Technique

Throughout this range, a basic technique of opinion presentation applies. Discussed below, this technique accommodates the principal purpose of editorial columning: to express an opinion intended to influence the thoughts and actions of others.

Like so many forms of writing, the editorial has three parts. Unlike some forms, notably the news story, the editorial cannot get along without one or two of its parts. (A news story can be a beginning, or lead sentence, alone; a magazine article may be devoid of a formal conclusion.)

The editorial's parts are introduction, discussion, and conclusion. The introduction identifies the topic and at least hints at the magazine's attitude (save strong assertions for the conclusion). The discussion describes the topic sufficiently to permit intelligent appraisal, asserts principles, and presents facts that will support the forthcoming conclusion. The conclusion then states, as tersely and unequivocally as possible, the magazine's wishes on the topic.

One desirable feature of this favored form of editorial is that it encourages positive thinking and constructive criticism. If the editorial is *against* a process or proposition, the act of writing something consciously called a conclusion impels the formulation of a desirable alternative—even if the alternative is to forget the process or proposition. If the editorial is *for* something, the conclusion makes the magazine's favorable opinion unmistakable.

Less-than-satisfactory editorials fall down in any of the three parts by either omitting or misusing them. The beginning that fails to identify the topic or suggest the stand that will be taken makes the reader ask, "What's he talking about?" or "Why is he telling me this?" The discussion that does not present pertinent propositions, principles, and facts in logical order will too soon send the reader to another page. Without the discussion, the editorial is a bald, unsupported statement of opinion—or worse, a witless waste of words. And what should a reader think of a conclusion such as, "This situation will bear watching," or "Something ought to be done"? Possibly what ought to be done is to publish a few guest editorials.

At the risk of presenting a less-than-superb example of the recommended form, here is an editorial favoring the message and mission of this book.

## EDITORIAL COMMENT

### Beginning

Writing about one's work can bring many satisfactions, not the least of which is found in the knowledge that an important professional obligation is being met.

### Discussion

Gratifications include the joy of seeing one's writing in print, greater esteem among fellow professionals, increased exposure, and promotion and employment opportunities. Writing can lead to lecture assignments, consulting, more writing, and tidy bits of extra cash.

The beauty of it all is that every one of these gratifications is a plus—pure profit from an act that of itself is the simple repayment of a debt. This debt, of course, is owed to the body of knowledge that, through education, continuing study, and work experience, enabled the writer to have something worth writing about.

Thus the practicing professional develops an object or concept that is useful to people within the reach of his voice or memorandum. On the supposition that his development also would be useful to many others similarly situated, he writes an article and gets its published. At this point he has added to the literature—the reservoir of knowledge from which he drew much of his proficiency. He has paid a professional debt.

## Conclusion

Clearly, then, an obligation to write arises in the natural course of professional development. Beyond the initial inertia, the burden of obligation is light and the rewards are dazzling.

# APPENDIX A: STYLE

Each magazine controls style, as it pertains to usages in spelling, capitalization, compounding, punctuation, numeralization, and abbreviation. The closer you adhere to a publisher's style, the easier your copy will be to edit. Although it is unlikely that you will be able to anticipate all style idiosyncrasies correctly, you can come close to most publications' practices by studying a standard. A comprehensive standard is the U.S. Government Printing Office Style Manual. Whole sections 3, 6, 9, and 11 of the 1967 Edition are reproduced on the following pages. Other useful publications are American Standard Abbreviations for Scientific and Engineering Terms (ASA Z 10.1-1941), published by the American Society of Mechanical Engineers, and USA and IEEE Standard Letter Symbols for Unit Used in Electrical Science and Electrical Engineering (USAS Y 10. 19; IEEE No. 260), published by the Institute of Electrical and Electronics Engineers.

# 3. CAPITALIZATION

**(See also Abbreviations; Guide to Capitalization)**

**3.1.** It is impossible to give rules that will cover every conceivable problem in capitalization. But by considering the purpose to be served and the underlying principles, it is possible to attain a considerable degree of uniformity. The list of approved forms given on pages 33 to 59 will serve as a guide. Manifestly such a list cannot be complete. The correct usage with respect to any term not included can be determined by analogy or by application of the rules.

## Proper names

**3.2.** Proper names are capitalized.

| | | |
|---|---|---|
| Rome | John Macadam | Italy |
| Brussels | Macadam family | Anglo-Saxon |

## Derivatives of proper names

**3.3.** Derivatives of proper names used with a proper meaning are capitalized.

| | | |
|---|---|---|
| Roman (of Rome) | Johannean | Italian |

**3.4.** Derivatives of proper names used with acquired independent common meaning, or no longer identified with such names, are lowercased. Since this depends upon general and long-continued usage, a more definite and all-inclusive rule cannot be formulated in advance. A list of derivatives is given on pages 41–42.

| | | |
|---|---|---|
| roman (type) | macadam (crushed rock) | italicize |
| brussels sprouts | watt (electric unit) | anglicize |
| venetian blinds | plaster of paris | pasteurize |

## Common nouns and adjectives in proper names

**3.5.** A common noun or adjective forming an essential part of a proper name is capitalized; the common noun used alone as a substitute for the name of a place or thing is not capitalized.

Massachusetts Avenue; the avenue
Washington Monument; the monument
Statue of Liberty; the statue
Hoover Dam; the dam
Boston Light; the light
Modoc National Forest; the national forest
Panama Canal; the canal
Soldiers' Home of Ohio; the soldiers' home
Johnson House (hotel); Johnson house (residence)
Crow Reservation; the reservation
Federal Express; the express
Cape of Good Hope; the cape
Jersey City; *also* Washington City; *but* city of Washington; the city
Cook County; the county
Great Lakes; the lakes
Lake of the Woods; the lake
North Platte River; the river
Lower California; *but* lower Mississippi
Charles the First; Charles I
Seventeenth Census; the 1960 census

**3.6.** If a common noun or adjective forming an essential part of a name becomes removed from the rest of the name by an intervening common noun or adjective, the entire expression is no longer a proper noun and is therefore not capitalized.

Union Station: union passenger station
Eastern States: eastern farming States
Western States: western farming States

**3.7.** A common noun used alone as a well-known short form of a specific proper name is capitalized.

the Capitol (at Washington); *but* State capitol
the Channel (English Channel)
the District (District of Columbia)
the Soldiers' Home (District of Columbia only)

**3.8.** The plural form of a common noun capitalized as part of a proper name is also capitalized.

Seventh and I Streets
Lakes Erie and Ontario
Potomac and James Rivers
State and Treasury Departments
British and French Governments
Presidents Washington and Adams

**3.9.** A common noun used with a date, number, or letter, merely to denote time or sequence, or for the purpose of reference, record, or temporary convenience, does not form a proper name and is therefore not capitalized. (See also rule 3.39, p. 29.)

abstract B
act of 1928
amendment 5
appendix C
article 1
book II
chapter III
chart B
class I
collection 6
column 2
drawing 6
exhibit D
figure 7
first district (not congressional)
form 4
graph 8
group 7
mile 7.5
page 2
paragraph 4
part I
plate IV
region 3
rule 8
schedule K
section 3
signature 4
station 27
table 4
title IV
treaty of 1919
volume X
war of 1914
ward 2

**3.10.** The following terms are lowercased, even with a name or number. (For capitalized forms, see geographic terms, pp. 45–46.)

aqueduct
breakwater
buoy
chute
dam (lowercase with number or in conjunction with lock; capitalize with name, *but* Boulder Dam site; Boulder Dam and site)
dike
dock
drydock
irrigation project
jetty
levee
lock
pier
reclamation project
ship canal
shipway
slip
spillway
tunnel (see also Tunnel, p. 57)
watershed
weir
wharf

## Definite article in proper names

**3.11.** To achieve greater distinction or to adhere to the authorized form, the word *the* (or its equivalent in a foreign language) used as a part of an official name or title is capitalized. When such name or title is used adjectively, *the* is not capitalized, nor is it supplied at any time when not in copy.

*British Consul* v. *The Mermaid* (title of legal case)
The Dalles (Oreg.); The Weirs (N.H.); *but* the Dalles region; the Weirs streets
The Hague; *but* the Hague Court; the Second Hague Conference
El Salvador; Las Cruces; L'Esterel
The Adjutant General (only when so in copy)

**3.12.** In common practice, rule 3.11 is disregarded in references to newspapers, periodicals, vessels, airships, trains, firm names, etc.

the Times
the Atlantic Monthly
the Washington Star
the *Mermaid*
the *Los Angeles*
the *U–3*
the Federal Express
the National Photo Co.
the Netherlands

## Particles in names of persons

**3.13.** In foreign names such particles as *d'*, *da*, *della*, *du*, *van*, and *von* are capitalized unless preceded by a forename or title. Individual usage, if ascertainable, should be followed.

Da Ponte; Cardinal da Ponte
Du Pont; E. I. du Pont de Nemours & Co.
Van Rensselaer; Stephen van Rensselaer
*but* d'Orbigny; Alcide d'Orbigny

**3.14.** In anglicized names such particles are usually capitalized, even if preceded by a forename or title, but individual usage, if ascertainable, should be followed.

Justice Van Devanter; Reginald De Koven
Thomas De Quincey; William De Morgan
Henry van Dyke (his usage)
Samuel F. Du Pont (his usage); Irénée du Pont
(for firm names, see p. 44)

**3.15.** If copy is not clear as to the form of such a name (for example, *La Forge* or *Laforge*), the two-word form should be used.

**3.16.** In names set in capitals, *de*, *von*, etc., are also capitalized.

## Names of organized bodies

**3.17.** The full names of existing or proposed organized bodies and their shortened names are capitalized; other substitutes, which are most often regarded as common nouns, are capitalized only in certain specified instances to indicate preeminence or distinction. (See list on pp. 33–59.)

National governmental units:
- U.S. Congress: 89th Congress; Congress; the Senate; the House; Committee of the Whole, the Committee; *but* committee (all other congressional committees)
- Department of Agriculture: the Department; Division of Publications, the Division; *similarly* all departmental units; *but* legislative, executive, and judicial departments
- Bureau of the Census: the Census Bureau, the Bureau
- Geological Survey: the Survey
- Interstate Commerce Commission: the Commission
- Government Printing Office: the Office
- Board of Commissioners of the District of Columbia: the Board of Commissioners; the Board
- American Embassy, British Embassy: the Embassy; *but* the consulate; the consulate general
- Treasury of the United States: General Treasury; National Treasury; Public Treasury; the Treasury; Treasury notes; New York Subtreasury, the subtreasury
- Department of Defense:
  - Military Establishment; Armed Forces; *but* armed services
  - U.S. Army: the Army; the Infantry; 81st Regiment; Army Establishment; the Army Band; Army officer; Regular Army officer; Reserve officer; Volunteer officer; *but* army shoe; Grant's army; Robinson's brigade; the brigade; the corps; the regiment; infantryman
  - U.S. Navy: the Navy; the Marine Corps; Navy (Naval) Establishment; Navy officer; *but* naval shipyard; naval officer; naval station
- French Ministry of Foreign Affairs; the Ministry; French Army; British Navy

International organizations:
- United Nations: the Council; the Assembly; the Secretariat
- Permanent Court of Arbitration: the Court; the Tribunal (only in the proceedings of a specific arbitration tribunal)
- Hague Peace Conference of 1907: the Hague Conference; the Peace Conference; the Conference

789–445°—67——3

**Common-noun substitutes:**

Virginia Assembly: the assembly; the senate; the house of delegates
California State Highway Commission: Highway Commission of California; the highway commission; the commission
Montgomery County Board of Health: the Board of Health, Montgomery County; the board of health; the board
Common Council of the City of Pittsburgh: the common council; the council
Buffalo Consumers' League: the consumers' league; the league
Republican Party: the party
Pennsylvania Railroad Co.: the Pennsylvania Railroad; Pennsylvania Co.; Pennsylvania Road; the railroad company; the company
Riggs National Bank: the Riggs Bank; the bank
Metropolitan Club: the club
Yale School of Law: Yale University School of Law; School of Law, Yale University; school of law

**3.18.** The names of members and adherents of organized bodies are capitalized to distinguish them from the same words used merely in a descriptive sense.

| | |
|---|---|
| a Representative (U.S. Congress) | a Socialist |
| a Republican | an Odd Fellow |
| an Elk | a Communist |
| a Liberal | a Boy Scout |
| a Shriner | a Knight (K.C., K.P., etc.) |

## Names of countries, domains, and administrative divisions

**3.19.** The official designations of countries, national domains, and their principal administrative divisions are capitalized only if used as part of proper names, as proper names, or as proper adjectives. (See table on p. 244.)

United States: the Republic; the Nation; the Union; the Government; *also* Federal, Federal Government; *but* republic (when not referring specifically to one such entity); republican (in general sense); a nation devoted to peace
New York State: the State, a State (a definite political subdivision of first rank); State of Veracruz; Balkan States; six States of Australia; State rights; *but* state (referring to a Federal Government, the body politic); foreign states; church and state; statehood; state's evidence
Territory (Canada): Yukon, Northwest Territories; the Territory(ies), Territorial; *but* territory of American Samoa, Guam, Virgin Islands
Ethiopian Empire: the Empire; *but* empire (in general sense)
Dominion of Canada: the Dominion; *but* dominion (in general sense)
Ontario Province, Province of Ontario: the Province, Provincial; *but* province, provincial (in general sense)
Crown Colony of Hong Kong, Cyprus: the colony, crown colony

**3.20.** The similar designations *commonwealth, confederation* (*federal*), *government, nation* (*national*), *powers, union,* etc., are capitalized only if used as part of proper names, as proper names, or as proper adjectives.

British Commonwealth, Commonwealth of Massachusetts: the Commonwealth; *but* commonwealth (in general sense)
Swiss Confederation: the Confederation; the Federal Council; the Federal Government; *but* confederation, federal (in general sense)
French Government: the Government; French and Italian Governments; Soviet Government; the Governments; *but* government (in general sense); the Churchill government; European governments
Cherokee Nation: the nation; *but* Greek nation; American nations
National Government (of any specific nation); *but* national customs
Allied Powers, Allies; *but* our allies, weaker allies (in World Wars I and II); Central Powers (in World War I); *but* the powers; European powers
Union of South Africa: the Union; *but* union (in general sense)

## Names of regions, localities, and geographic features

**3.21.** A descriptive term used to denote a definite region, locality, or geographic feature is a proper name and is therefore capitalized; also for temporary distinction a coined name of a region is capitalized.

the North Atlantic States; the Gulf States; the Central States; the Pacific Coast States; the Lake States; East North Central States; Eastern North Central States; Far Western States; Eastern United States
the West; the Midwest; the Middle West; Far West
the Eastern Shore (Chesapeake Bay)
the Badlands (S. Dak. and Nebr.)
the Continental Divide (Rocky Mountains)
Deep South; Midsouth
the Occident; the Orient
the Far East; Far Eastern; the East
Middle East, Middle Eastern, Mideast, Mideastern (Asia)
Near East (Balkans, etc.)
the Promised Land
the Continent (continental Europe)
the Western Hemisphere
the North Pole; the North and South Poles
the Temperate Zone; the Torrid Zone
the East Side (section of a city)
the Driftless Area (Mississippi Valley)
Western Germany; Western Europe (political entities)

**3.22.** A descriptive term used to denote mere direction or position is not a proper name and is therefore not capitalized.

north; south; east; west
northerly; northern; northward
eastern; oriental; occidental
east Pennsylvania; southern California
west Florida; *but* West Florida (1763–1819)
eastern region; western region
north-central region
east coast; eastern seaboard
central Europe; south Germany; southern France
*but* East Germany; West Germany (political entities)

## Names of calendar divisions

**3.23.** The names of divisions are capitalized.

January; February; March; etc.
Monday; Tuesday; Wednesday; etc.
*but* spring; summer; autumn (fall); winter

## Names of historic events, etc.

**3.24.** The names of holidays, ecclesiastic feast and fast days, and historic events are capitalized.

Battle of Bunker Hill
Battle of the Giants
Christian Era; Middle Ages; *but* 20th century
Feast of the Passover; the Passover
Fourth of July; the Fourth
Reformation
Renaissance
Veterans Day
War of 1812; World War II; *but* war of 1914; Korean war

## Trade names

**3.25.** Trade names, variety names, and names of market grades and brands are capitalized. Common nouns following such names are not capitalized. (See market grades, p. 48; trade names, pp. 56, 277.)

Foamite (trade name)
Plexiglas (trade name)
Snow Crop (trade name)
Choice lamb (market grade)
Yellow Stained cotton (market grade)
Red Radiance rose (variety)

## Scientific names

**3.26.** The name of a phylum, class, order, family, or genus is capitalized; the name of a species is not capitalized, even though derived from a proper name.

Arthropoda (phylum), Crustacea (class), Hypoparia (order), Agnostidae (family), *Agnostus* (genus)
*Agnostus canadensis; Aconitum wilsoni; Epigaea repens* (genus and species)

**3.27.** In scientific descriptions coined terms derived from proper names are not capitalized.

aviculoid menodontine

**3.28.** A plural formed by adding *s* to a Latin generic name is capitalized.

Rhynchonellas Spirifers

**3.29.** In soil science the 24 soil classifications are capitalized. (For complete list, see p. 54.)

Alpine Meadow Bog Brown

**3.30.** The words *sun, moon,* and *earth* are capitalized only if used in association with the names of other astronomical bodies that are capitalized.

The nine known planets, in the order of distance from the Sun, are Mercury, Venus, the Earth, Mars, Jupiter, Saturn, Uranus, Neptune, and Pluto.

**3.31.** For lists of geologic and physiographic terms, see page 241.

## Fanciful appellations

**3.32.** A fanciful appellation used with or for a proper name is capitalized.

the Big Four
the Dust Bowl
the Great Society
the Hub
the Keystone State
the New Deal
the New Frontier
the Pretender

## Personification

**3.33.** A vivid personification is capitalized.

The Chair recognized the gentleman from New York.
For Nature wields her scepter mercilessly.
All are architects of Fate,
Working in these walls of Time.

## Religious terms

**3.34.** All words denoting the Deity except *who, whose,* and *whom;* all names for the Bible and other sacred writings; and all names of confessions of faith and of religious bodies and their adherents and words specifically denoting Satan are capitalized.

Heavenly Father; the Almighty; Thee; Thou; He; Him; *but* himself; [God's] fatherhood
Mass; red Mass; Communion
Divine Father; *but* divine providence; divine guidance; divine service
Son of Man; Jesus' sonship; the Messiah; *but* a messiah; messiahship; messianic; messianize; christology; christological
Bible, Holy Scriptures, Scriptures; Koran; *also* Biblical; Scriptural; Koranic
Gospel (memoir of Christ); *but* gospel truth
Apostles' Creed; Augsburg Confession; Thirty-nine Articles
Episcopal Church: an Episcopalian; Catholicism; a Protestant
Christian; *also* Christendom; Christianity; Christianize
Black Friars; Brother(s); King's Daughters; Daughter(s); Ursuline Sisters; Sister(s)
Satan; His Satanic Majesty; Father of Lies; the Devil; *but* a devil; the devils; devil's advocate

## Titles of persons

**3.35.** Any title immediately preceding a name is capitalized.

President Roosevelt
King George
Ambassador Gibson
Lieutenant Fowler
Chairman Smith
Nurse Cavell
Professor Leverett
Examiner Jones
*but* vice-presidential candidate Humphrey
baseball player Mantle
maintenance man Jones

**3.36.** To indicate preeminence or distinction in certain specified instances, a common-noun title immediately following the name of a person or used alone as a substitute for it is capitalized.

Title of a head or assistant head of state:

Lyndon B. Johnson, President of the United States: the President; the President-elect; the Executive; the Chief Magistrate; the Commander in Chief; ex-President Eisenhower; former President Truman; *similarly* the Vice President; the Vice-President-elect; ex-Vice-President Nixon

Harry W. Nice, Governor of Maryland: the Governor of Maryland; the Governor; *similarly* the Lieutenant Governor; *but* secretary of state of Idaho; attorney general of Maine

Title of a head or assistant head of an existing or proposed National or District governmental unit:

Dean Rusk, Secretary of State: the Secretary; *similarly* the Acting Secretary; the Under Secretary; the Assistant Secretary; the Director; the Chief or Assistant Chief; the Chief Clerk; etc.; *but* Secretaries of the military departments; secretaryship

Titles of the military:

General of the Army(ies): United States only; Supreme Allied Commander; Gen. Omar N. Bradley, Chairman, Joint Chiefs of Staff; Joint Chiefs of Staff; Chief of Staff, U.S. Air Force; the Chief of Staff; *but* the general (military title standing alone not capitalized)

Titles of members of diplomatic corps:

Walter S. Gifford, Ambassador Extraordinary and Plenipotentiary: the American Ambassador; the British Ambassador; the Ambassador; the Senior Ambassador; His Excellency; *similarly* the Envoy Extraordinary and Minister Plenipotentiary; the Envoy; the Minister; the Chargé d'Affaires; the Chargé; Ambassador at Large; Minister Without Portfolio; *but* the consul general; the consul; the attaché; etc.

Title of a ruler or prince:

Elizabeth II, Queen of England: the Queen; the Crown; Her Most Gracious Majesty; Her Majesty; *similarly* the Emperor; the Sultan; etc.

Edward, Prince of Wales: the Prince; His Royal Highness

Titles not capitalized:

Charles F. Hughes, rear admiral, U.S. Navy: the rear admiral

Cloyd H. Marvin, president of George Washington University: the president

C. H. Eckles, professor of dairy husbandry: the professor

John Smith, chairman of the committee: the chairman

**3.37.** In formal lists of delegates and representatives of governments, all titles and descriptive designations immediately following the names should be capitalized if any one is capitalized.

**3.38.** A title in the second person is capitalized.

| | | |
|---|---|---|
| Your Excellency | Mr. Chairman | Not salutation: |
| Your Highness | Mr. Secretary | my dear General |
| Your Honor | | my dear sir |

## Titles of publications, papers, documents, acts, laws, etc.

**3.39.** In the full or short English titles of periodicals, series of publications, annual reports, historic documents, and works of art, the first word and all important words are capitalized.

Statutes at Large; Revised Statutes; District Code; Bancroft's History; Journal (House or Senate) (short titles); *but* the code; the statutes

Atlantic Charter; Balfour Declaration; *but* British white paper

American Journal of Science

Saturday Evening Post; the Post

Philadelphia Inquirer

Chicago's American; *but* Chicago American Publishing Co.

Reader's Digest; *but* New York Times Magazine; Newsweek magazine

Monograph 55; Research Paper 123; Bulletin 420; Circular A; Article 15, Uniform Code of Military Justice; Senate Document 70; House Resolution 45; Presidential Proclamation No. 24; Executive Order No. 24; Royal Decree No. 24; Public Law 89–1; Private and Union Calendars; Calendar No. 80; Calendar Wednesday; Committee Print No. 32, committee print; *but* Senate bill 416; House bill 61

Annual Report of the Public Printer, 1966; *but* seventh annual report, 19th annual report (see rule 11.9, p. 171)

Declaration of Independence; the Declaration

Constitution (United States or with name of country); constitutional; *but* New York State constitution; first amendment, 12th amendment (see rule 11.9, p. 171)

Kellogg Pact; North Atlantic Pact; Atlantic Pact; Treaty of Versailles; Jay Treaty; *but* treaty of peace, the treaty (descriptive designations); treaty of 1919

*United States* v. *Four Hundred Twenty-two Casks of Wine* (legal case) (see also rule 18.33, p. 231)

The Blue Boy (painting)

**3.40.** All principal words are capitalized in titles of addresses, articles, books, captions, chapter and part headings, editorials, essays, headings, headlines, motion pictures and plays (including TV and radio programs), papers, short poems, reports, songs, subheadings, subjects, and themes. The foregoing are also quoted. (See rule 9.118, p. 148, for examples of capitalization and use of quotation marks.)

**3.41.** In the short or popular titles of acts (Federal, State, or foreign) the first word and all important words are capitalized.

Revenue Act; Walsh-Healey Act; Panama Canal Act; Classification Act; *but* revenue act(s); act of 1926, 1926 act; the act; Harrison narcotic law; Harrison narcotic bill; interstate commerce law

**3.42.** The capitalization of the titles of books, etc., written in a foreign language is to conform to national practice in that language. For further details and examples, see section on foreign languages.

**3.43.** In lists, including bibliographies and synonymies, and in footnote citations, capitalization will conform to the rules of this chapter, unless the work requires its own established style.

### First words

**3.44.** The first word of a sentence, of an independent clause or phrase, of a direct quotation, of a line of poetry, or of a formally introduced series of items or phrases following a comma or colon is capitalized.

The question is, Shall the bill pass?
He asked, "And where are you going?"

Lives of great men all remind us
We can make our lives sublime.

The vote was as follows: In the affirmative, 23; in the negative, 11; not voting, three.

**3.45.** The first word of a fragmentary quotation is not capitalized.

He objected "to the phraseology, not to the ideas."

**3.46.** The first word following a colon, an exclamation point, or an interrogation point is not capitalized if the matter following is merely a supplementary remark making the meaning clearer.

Revolutions are not made: they come.
Intelligence is not replaced by mechanism: even the televox must be guided by its master's voice.
But two months dead! nay, not so much; not two.

What is this?
Your knees to me? to your corrected son?

**3.47.** The first word following *Whereas* in resolutions, contracts, etc., is not capitalized; the first word following an enacting or resolving clause is capitalized.

Whereas the Constitution provides * * *; and
Whereas Congress has passed a law * * *;
Whereas, moreover, * * *: Therefore be it
Whereas the Senate provided for the * * *: Now, therefore, be it
*Resolved*, That * * *; and be it further
*Resolved (jointly)*, That * * *
*Resolved by the House of Representatives (the Senate concurring)*, That * * *. (Concurrent resolution, Federal Government.)
*Resolved by the Senate of Oklahoma (the House of Representatives concurring therein)*, That * * *. (Concurrent resolution, using name of State.)
*Resolved by the senate (the house of representatives concurring therein)*, That * * *. (Concurrent resolution, not using name of State.)
*Resolved by the Assembly and Senate of the State of California (jointly)*, That * * *. (Joint resolution, using name of State.)
*Resolved by the Washington Board of Trade*, That * * *
*Provided*, That * * *
*Provided further*, That * * *
*Provided, however*, That * * *
*And provided further*, That * * *
*Ordered*, That * * *
*Be it enacted*, That * * *

## Center and side heads

**3.48.** Unless otherwise marked, (1) centerheads are set in capitals, and (2) sideheads are set in lowercase and only the first word and proper names are capitalized. In centerheads making two lines, wordbreaks should be avoided. The first line should be centered and set as full as possible, but it is not to be set to fill the measure by unduly wide spacing.

**3.49.** Except as indicated elsewhere, everything in a cap heading is set in caps; in a cap and small-cap heading, in caps and small caps; and in a small-cap heading, in small caps, including, if available, parentheses, brackets, and figures. En quads are used between words.

**3.50.** In heads set in caps, a small-cap *c* or *ac*, if available, is used in such names as *McLean* or *MacLeod;* otherwise a lowercase *c* or *ac* is used. In heads set in small caps, an apostrophe is used instead of the *c*, but a space is used after the *ac*.

**3.51.** In such names as *LeRoy*, *DeHostis*, *LaFollette*, etc. (one-word forms only), set in caps, the second letter of the particle is made a small cap, if available; otherwise lowercase is used. In heads set in small caps, a space is used.

**3.52.** In matter set in caps and small caps or caps and lowercase, capitalize all principal words, including parts of compounds which would be capitalized standing alone. The articles *a*, *an*, and *the;* the prepositions *at*, *by*, *for*, *in*, *of*, *on*, *to*, and *up;* the conjunctions *and*, *as*, *but*, *if*, *or*, and *nor;* and the second element of a compound numeral are not capitalized. (See also rule 9.118, p. 148.)

Airplanes Versus Battleships
World in All-Out War
Man Hit With 2-Inch Pipe
No-Par-Value Stock for Sale
Price-Cutting War
Yankees May Be Winners
Ex-Senator Is To Be Admitted
Notice of Filing and Order on Exemption From Requirements

*but* Building on Twenty-first Street (if spelled)
One Hundred and Twenty-three Years (if spelled)
Only One-tenth of Shipping Was Idle
Many 35-Millimeter Films in Production
Built-Up Stockpiles Are Necessary (*Up* is adverb here)

**3.53.** *Continued* heads will be set according to rules 14.51–14.53, pages 189–190.

**3.54.** If a normally lowercased short word is used in juxtaposition with a capitalized word of like significance in the sentence, it should also be capitalized.

Buildings In and Near Minneapolis

**3.55.** In a heading set in caps and lowercase or in caps and small caps, a normally lowercased last word, if it is the only lowercased word in the heading, should also be capitalized.

All Returns Are In

**3.56.** The first element of an infinitive is capitalized.

Controls To Be Applied *but* Aid Sent to Disaster Area

**3.57.** In matter set in caps and small caps, the abbreviations *etc.* and *et al.* are set in small caps; in matter set in caps and lowercase, these abbreviations are set in lowercase.

PLANES, GUNS, SHIPS, ETC. Planes, Guns, Ships, etc.
JAMES BROS. ET AL. James Bros. et al.

**3.58.** As accents in cap lines have a tendency to break off in proofing, presswork, etc., they may be omitted, even if the same words carry accents in text.

**3.59.** Paragraph series letters in parentheses appearing in heads set in caps, caps and small caps, small caps, or in caps and lowercase are to be set as in copy.

### Addresses, salutations, and signatures

**3.60.** The first word and all principal words in addresses, salutations, and signatures are capitalized. (See "Datelines, Addresses, and Signatures," p. 221.)

### Interjections

**3.61.** The interjection *O* is always capitalized; within a sentence other interjections are not capitalized.

Sail on, O Ship of State!
For lo! the days are hastening on.
But, oh, how fortunate!

### Historic or documentary accuracy

**3.62.** Where historic or documentary accuracy is required, capitalization and other features of style of the original text should be followed.

# 6. COMPOUND WORDS

(See also Guide to Compounding; Word Division (supplement to STYLE MANUAL), description on p. 2)

**6.1.** A compound word is a union of two or more words, either with or without a hyphen. It conveys a unit idea that is not as clearly or quickly conveyed by the component words in unconnected succession. The hyphen in a compound is a mark of punctuation that not only unites but separates the component words, and thus facilitates understanding, aids readability, and insures correct pronunciation.

**6.2.** In applying the following rules and in using the Guide to Compounding, the living fluidity of our language should be kept in mind. Word forms constantly undergo modification. Two-word forms first acquire the hyphen, later are printed as one word, and not infrequently the transition is from the two- to the one-word form, bypassing the hyphen stage.

**6.3.** The rules as laid down cannot be applied inflexibly. Exceptions must necessarily be allowed, so that general good form will not be offended. However, current language trends point definitely to closing up words which, through frequent use, have become associated in the reader's mind as units of thought. The tendency to amalgamate words, particularly two short words, assures easier continuity, and is a natural progression from the older and less flexible treatment of words.

## General rules

**6.4.** In general, omit the hyphen when words appear in regular order and the omission causes no ambiguity in sense or sound. (See also rule 6.16, p. 75.)

banking hours
blood pressure
book value
census taker
day laborer
eye opener
fellow citizen
living costs
palm oil
patent right
real estate
rock candy
training ship
violin teacher

**6.5.** Compound two or more words to express a literal or nonliteral (figurative) unit idea that would not be as clearly expressed in unconnected succession.

afterglow
bookkeeping
cupboard
forget-me-not
gentleman
newsprint
right-of-way
whitewash

**6.6.** Unless otherwise indicated, a derivative of a compound retains the solid or hyphened form of the original compound.

coldbloodedness
footnoting
ill-advisedly
outlawry
praiseworthiness
railroader
X-rayer
Y-shaped

**6.7.** Except after the short prefixes *co*, *de*, *pre*, *pro*, and *re*, which are generally printed solid, a hyphen is used to avoid doubling a vowel or tripling a consonant. (See also rules 6.29, 6.32, p. 77.)

cooperation
deemphasis
preexisting
anti-inflation
micro-organism
semi-independent
brass-smith
Inverness-shire
thimble-eye
ultra-atomic
shell-like
hull-less



789-445°—67——6

## Solid compounds

**6.8.** Print solid two nouns that form a third when the compound has only one primary accent, especially when the prefixed noun consists of only one syllable or when one of the elements loses its original accent.

| | | |
|---|---|---|
| airship | cupboard | footnote |
| bathroom | dressmaker | locksmith |
| bookseller | fishmonger | workman |

**6.9.** Print solid a noun consisting of a short verb and an adverb as its second element, except when the use of the solid form would interfere with comprehension.

| | | | |
|---|---|---|---|
| blowout | hangover | pickup | throwaway |
| breakdown | holdup | runoff | *but* cut-in |
| flareback | makeready | setup | run-in |
| giveaway | markoff | showdown | tie-in |

**6.10.** Compounds beginning with the following nouns are usually printed solid.

| | | | |
|---|---|---|---|
| book | house | school | way |
| eye | mill | shop | wood |
| horse | play | snow | work |

**6.11.** Compounds ending in the following are usually printed solid, especially when the prefixed word consists of one syllable. (See also rules 8.5, p. 131; 8.7, p. 135.)

| | | | |
|---|---|---|---|
| berry | house | piece | weed |
| blossom | keeper | power | wide |
| boat | keeping | proof | wise |
| book | light | room | woman |
| borne | like | shop | wood |
| bound | maker | smith | work |
| brained | making | stone | worker |
| bush | man | store | working |
| fish | master | tail | worm |
| flower | mate | tight | wort |
| grower | mill | time (not clock) | writer |
| hearted | mistress | ward | writing |
| holder | monger | way | yard |

**6.12.** Print solid *any, every, no,* and *some* when combined with *body, thing,* and *where;* when *one* is the second element, print as two words if meaning a single or particular person or thing; to avoid mispronunciation, print *no one* as two words at all times.

| | | | |
|---|---|---|---|
| anybody | everybody | nobody | somebody |
| anything | everything | nothing | something |
| anywhere | everywhere | nowhere | somewhere |
| anyone | everyone | no one | someone |

*but* any one of us may stay; every one of the pilots is responsible.

**6.13.** Print as one word compound personal pronouns.

| | | |
|---|---|---|
| herself | oneself | thyself |
| himself | ourselves | yourself |
| itself | themselves | yourselves |
| myself | | |

**6.14.** Print as one word compass directions consisting of two points, but use a hyphen after the first point when three points are combined.

| | |
|---|---|
| northeast | north-northeast |
| southwest | south-southwest |

## Unit modifiers

(See also rule 9.58, p. 142.)

**6.15.** Print a hyphen between words, or abbreviations and words, combined to form a unit modifier immediately preceding the word modified, except as indicated in rule 6.16 and elsewhere throughout this chapter. This applies particularly to combinations in which one element is a present or past participle.

Baltimore-Washington road
collective-bargaining talks
contested-election case
contract-bar rule
drought-stricken area
English-speaking nation
fire-tested material
Federal-State-local cooperation
German-English descent
guided-missile program
hard-of-hearing class
high-speed line
large-scale project
law-abiding citizen
long-term loan
long-term-payment loan
lump-sum payment
most-favored-nation clause
multiple-purpose uses
no-par-value stock
part-time personnel
rust-resistant covering
service-connected disability
tool-and-die maker
1-inch diameter; 2-inch-diameter pipe
10-word telegram
a 4-percent increase; *but* 4 percent [of] hydrochloric acid, 4 percent [of] interest
U.S.-owned property; U.S.-flag ship

**6.16.** Where meaning is clear and readability is not aided, it is not necessary to use a hyphen to form a temporary or made compound. Restraint should be exercised in forming unnecessary combinations of words used in normal sequence.

atomic energy power
bituminous coal industry
child welfare plan
civil rights case
civil service examination
durable goods industry
flood control study
free enterprise system
high school student; elementary school grade
income tax form
interstate commerce law
land bank loan
land use program
life insurance company
mutual security funds
national defense appropriation
natural gas company
per capita expenditure
portland cement plant
production credit loan
public utility plant
real estate tax
small businessman
social security pension
soil conservation measures
special delivery mail; parcel post delivery
speech correction class
*but* no-hyphen rule (readability aided); *not* no hyphen rule

**6.17.** Print without a hyphen a compound predicate adjective or predicate noun the second element of which is a present participle.

The duties were price fixing.
The effects were far reaching.
The shale was oil bearing.
The area was used for beet raising.

**6.18.** Print without a hyphen a compound predicate adjective the second element of which is a past participle; also, omit the hyphen in a predicate modifier of comparative or superlative degree.

The area is drought stricken.
The paper is fine grained.
The boy is freckle faced.
This material is fire tested.
The cars are higher priced.
The reporters are best informed.

**6.19.** Print without a hyphen a two-word modifier the first element of which is a comparative or superlative.

better drained soil
best liked books
higher level decision
highest priced apartment
larger sized dress
better paying job
lower income group
*but* uppercrust society
lowercase, uppercase type (printing)
undercoverman
upperclassman
bestseller (noun)
lighter-than-air craft
higher-than-market price

**6.20.** Do not use a hyphen in a two-word unit modifier the first element of which is an adverb ending in *ly*, nor use hyphens in a three-word unit modifier the first two elements of which are adverbs.

eagerly awaited moment
heavily laden ship
unusually well preserved specimen
very well defined usage
longer than usual lunch period
very well worth reading
not too distant future
often heard phrase
*but* ever-normal granary
ever-rising flood
still-new car
still-lingering doubt
well-known lawyer
well-kept farm

**6.21.** Proper nouns used as unit modifiers, either in their basic or derived form, retain their original form; but the hyphen is printed after combining forms.

Latin American countries
North Carolina roads
South American trade
United States laws
Red Cross nurse
Winston-Salem festival
Washington-Wilkes-Barre route
Afro-American program
Anglo-Saxon period
Franco-Prussian War
*but* Indochina[1] border
Minneapolis-St. Paul region
North American-South American sphere
French-English descent

**6.22.** Do not confuse a modifier with the word it modifies.

elderly clothesman
competent shoemaker
field canning factory
gallant serviceman
light blue hat
average taxpayer
American flagship
well-trained schoolteacher
*but* common stockholder; small businessman
old-clothes man
wooden-shoe maker
tomato-canning factory
service men and women
light-blue hat
income-tax payer
American-flag ship
elementary school teacher

**6.23.** Where two or more hyphened compounds have a common basic element and this element is omitted in all but the last term, the hyphens are retained.

2- or 3-em quads, *not* 2 or 3-em quads; 2- to 3- and 4- to 5-ton trucks
2- by 4-inch boards, *but* 2 to 6 inches wide
8-, 10-, and 16-foot boards
moss- and ivy-covered walls, *not* moss and ivy-covered walls
long- and short-term money rates, *not* long and short-term money rates
*but* twofold or threefold, *not* two or threefold
goat, sheep, and calf skins, *not* goat, sheep, and calfskins
intrastate and intracity, *not* intra-state and -city
American owned and managed companies
preoperative and postoperative examination

**6.24.** Do not use a hyphen in a unit modifier consisting of a foreign phrase.

ante bellum days
bona fide transaction
ex officio member
per capita tax
per diem employee
prima facie evidence

**6.25.** Do not print a hyphen in a unit modifier containing a letter or a numeral as its second element.

abstract B pages
article 3 provisions
class II railroad
grade A milk
point 4 program
ward D beds

**6.26.** Do not use a hyphen in a unit modifier enclosed in quotation marks unless it is normally a hyphened term, but quotation marks are not to be used in lieu of a hyphen. (See also rule 9.123, p. 149.)

"blue sky" law "good neighbor" policy "tie-in" sale *but* right-to-work law

[1] Decision of U.S. Board on Geographic Names.

**6.27.** Print combination color terms as separate words, but use hyphen when such color terms are unit modifiers.

bluish green
dark green
orange red
bluish-green feathers
iron-gray sink
silver-gray body

**6.28.** Do not use a hyphen between independent adjectives preceding a noun.

big gray cat
a fine old southern gentleman

## Prefixes, suffixes, and combining forms

**6.29.** Print solid combining forms and prefixes, except as indicated elsewhere.

*after*birth
*Anglo*mania
*ante*date
*anti*slavery
*bi*weekly
*by*law
*circum*navigation
*cis*alpine
*co*operate
*contra*position
*counter*case
*de*energize
*demi*tasse
*ex*communicate
*extra*curricular
*fore*tell
*heroi*comic
*hyper*sensitive
*hypo*acid
*in*bound
*infra*red
*inter*view
*intra*spinal
*intro*vert
*iso*metric
*macro*analysis
*meso*thorax
*meta*genesis
*micro*phone
*mis*state
*mono*gram
*multi*color
*neo*phyte
*non*neutral
*off*set
*out*bake
*over*active
*pan*cosmic
*para*centric
*parti*coated
*peri*patetic
*plano*convex
*poly*nodal
*post*script
*pre*exist
*pro*consul
*pseudo*scholastic
*re*enact
*retro*spect
*semi*official
*step*father
*sub*secretary
*super*market
*thermo*couple
*trans*onic
*trans*ship
*tri*color
*ultra*violet
*un*necessary
*under*flow

**6.30.** Print solid combining forms and suffixes, except as indicated elsewhere.

port*able*
cover*age*
oper*ate*
plebis*cite*
twenty*fold*
spoon*ful*
kilo*gram*
geo*graphy*
man*hood*
self*ish*
pump*kin*
meat*less*
out*let*
wave*like*
procure*ment*
inner*most*
partner*ship*
lone*some*
home*stead*
north*ward*
clock*wise*

**6.31.** Print solid words ending in *like*, but use a hyphen to avoid tripling a consonant or when the first element is a proper name.

lifelike
lilylike
bell-like
girllike
Florida-like
Truman-like

**6.32.** Use a hyphen or hyphens to prevent mispronunciation, to insure a definite accent on each element of the compound, or to avoid ambiguity.

anti-hog-cholera serum
co-op
mid-ice
non-civil-service position
non-tumor-bearing tissue
re-cover (cover again)
re-sorting (sort again)
re-treat (treat again)
un-ionized
un-uniformity

**6.33.** Use a hyphen to join duplicated prefixes.

re-redirect
sub-subcommittee
super-superlative

**6.34.** Print with a hyphen the prefixes *ex*, *self*, and *quasi*.

ex-governor
ex-serviceman
ex-trader
ex-vice-president
self-control
self-educated
*but* selfhood
selfsame
quasi-academic
quasi-argument
quasi-corporation
quasi-young

**6.35.** Unless usage demands otherwise, use a hyphen to join a prefix or combining form to a capitalized word. (The hyphen is retained in words of this class set in caps.)

anti-Arab
pro-British
un-American
non-Government
post-World War II *or* post-Second World War

*but* nongovernmental
overanglicize
prezeppelin
transatlantic

## Numerical compounds

**6.36.** Print a hyphen between the elements of compound numbers from twenty-one to ninety-nine and in adjective compounds with a numerical first element. (See also rule 11.23, p. 173.)

twenty-one
twenty-first
6-footer
24-inch ruler
3-week vacation
8-hour day
10-minute delay
20th-century progress

3-to-1 ratio
5-to-4 vote
.22-caliber cartridge
2-cent-per-pound tax
four-in-hand tie
three-and-twenty
two-sided question
multimillion-dollar fund

*but* one hundred and twenty-one
100-odd
foursome
threescore
foursquare
$20 million airfield

**6.37.** Print without a hyphen a modifier consisting of a possessive noun preceded by a numeral. (See also rule 5.31, p. 71.)

1 month's layoff
1 week's pay

2 hours' work
3 weeks' vacation

**6.38.** Print a hyphen between the elements of a fraction, but omit it between the numerator and the denominator when the hyphen appears in either or in both.

one-thousandth
two-thirds
two one-thousandths

twenty-three thirtieths
twenty-one thirty-seconds
three-fourths of an inch

**6.39.** A unit modifier following and reading back to the word or words modified takes a hyphen and is always printed in the singular.

motor, alternating-current, 3-phase, 60-cycle, 115-volt
glass jars: 5-gallon, 2-gallon, 1-quart
belts: 2-inch, 1¼-inch, ½-inch, ¼-inch

## Civil and military titles

**6.40.** Do not hyphen a civil or military title denoting a single office, but print a double title with a hyphen. (See also rule 5.6, p. 66.)

ambassador at large
assistant attorney general
commander in chief
comptroller general
Congressman at Large
major general
sergeant at arms

notary public
secretary general
under secretary; *but* under-secretaryship
vice president;[2] *but* vice-presidency
secretary-treasurer
treasurer-manager

**6.41.** The adjectives *elect* and *designate*, as the last element of a title, require a hyphen.

President-elect
Vice-President-elect

ambassador-designate
minister-designate

---

[2] In official usage, the title of Vice President of the United States is written without a hyphen; the hyphen is also omitted in all like titles, such as vice admiral, vice consul, etc.

## Scientific and technical terms

**6.42.** Do not print a hyphen in scientific terms (names of chemicals, diseases, animals, insects, plants) used as unit modifiers if no hyphen appears in their original form. (See list of plant names, p. 131, and insect names, p. 135.)

carbon monoxide poisoning
guinea pig raising
hog cholera serum
methyl bromide solution
stem rust control
whooping cough remedy
*but* screw-worm raising
Russian-olive plantings
white-pine weevil
Douglas-fir tree

**6.43.** Chemical elements used in combination with figures use a hyphen, except with superior figures.

polonium-210 uranium-235; *but* $U^{235}$; $Sr^{90}$; ${}_{92}U^{234}$ Freon-12

**6.44.** Note use of hyphens and closeup punctuation in chemical formulas.

9-nitroanthra(1,9,4,10)bis(1)oxathiazone-2,7-bisdioxide
Cr-Ni-Mo
2,4-D

**6.45.** Print a hyphen between the elements of technical compound units of measurement.

candle-hour
horsepower-hour
kilowatt-hour
light-year
passenger-mile

## Improvised compounds

**6.46.** Print with a hyphen the elements of an improvised compound.

blue-pencil (v.)
18-year-old (n.)
first-come-first-served basis
know-it-all (n.)
know-how (n.)
make-believe (n.)
stick-in-the-mud (n.)
let-George-do-it attitude
how-to-be-beautiful course
hard-and-fast rule

**6.47.** Use hyphens in a prepositional-phrase compound noun consisting of three or more words.

cat-o'-nine-tails
government-in-exile
grant-in-aid
jack-in-the-box
man-of-war
mother-in-law
mother-of-pearl
patent-in-fee
*but* coat of arms
heir at law
next of kin
officer in charge

**6.48.** When the corresponding noun form is printed as separate words, the verb is always hyphened.

cold-shoulder blue-pencil cross-brace

**6.49.** Print a hyphen in a compound formed of repetitive or conflicting terms and in a compound naming the same thing under two aspects.

boogie-woogie
comedy-ballet
dead-alive
devil-devil
farce-melodrama
pitter-patter
walkie-talkie
willy-nilly
young-old

**6.50.** Use a hyphen in a nonliteral compound expression containing an apostrophe in its first element.

asses'-eyes
ass's-foot
bull's-eye
cat's-paw
crow's-nest
*but* The cat's paw is soft.
There is the crow's nest.

**6.51.** Print a hyphen to join a single capital letter to a noun or a participle.

| | | |
|---|---|---|
| H-bomb | U-boat | X-raying |
| I-beam | V-necked | S-iron |
| T-shaped | X-ray | T-square |

**6.52.** Print idiomatic phrases without hyphens.

| | |
|---|---|
| come by | insofar as |
| inasmuch as | Monday week |

# 9. PUNCTUATION

**9.1.** Punctuation is a device to clarify the meaning of written or printed language. Well-planned word order requires a minimum of punctuation. The trend toward less punctuation calls for skillful phrasing to avoid ambiguity and to insure exact interpretation. The MANUAL can only offer general rules of text treatment. A rigid design or pattern of punctuation cannot be laid down, except in broad terms. The adopted style, however, must be consistent and be based on sentence structure.

**9.2.** The general principles governing the use of punctuation are (1) that if it does not clarify the text it should be omitted, and (2) that in the choice and placing of punctuation marks the sole aim should be to bring out more clearly the author's thought. Punctuation should aid in reading and prevent misreading.

## Apostrophe

(See "Possessives and apostrophes," pp. 70–71.)

## Brace

**9.3.** The brace is used to show the relation of one line or group of lines to another group of lines. The point of the brace is placed toward the fewer number of lines; or if the number of lines is the same, toward the single group. For examples of braces used in tabular matter, see rule 14.35, page 187.

| | | | | |
|---|---|---|---|---|
| Supervision of timber sales. | 1-hour jobs | District 1<br>District 7 | 1½ hours' travel------ | Sales conducted monthly from May to July. |
| | 2-hour jobs | District 6<br>District 4 | 1 hour's travel-------- | |
| | 3-hour jobs | District 2<br>District 3<br>District 5 | 2 hours' travel-------- | |

## Brackets

***Brackets, in pairs, are used—***

**9.4.** To indicate a correction, a supplied omission, an interpolation, a comment, or a caution that an error is reproduced literally. (For use of parentheses, see rule 9.80, p. 144.)

He came on the 3d [2d] of July.
Our conference [lasted] 2 hours.
The general [Washington] ordered him to leave.
The paper was as follows [reads]:
I do not know. [Continues reading:]
[Chorus of "Mr. Chairman."]
They fooled only themselves. [Laughter.]
Our party will always serve the people [applause] in spite of the opposition [loud applause]. (If more than one bracketed interpolation, both are included within the sentence.)
The WITNESS. He did it that way [indicating].
Q. Do you know these men [handing witness a list]?
The bill had *not* been paid. [Italic added.] *or* [Emphasis added.]
The statue [sic] was on the statute books.
The WITNESS. This matter is classified. [Deleted.]
[Deleted.]
Mr. JONES. Hold up your hands. [Show of hands.]
Answer [after examining list]. Yes; I do.
Q. [Continuing.]
A. [Reads:]
A. [Interrupting.]



789-445°—67——10

**9.5.** In bills, contracts, etc., to indicate matter that is to be omitted.

**9.6.** In mathematics, to denote that enclosed matter is to be treated as a unit. (For examples, see p. 178.)

**9.7.** A single bracket may be used to indicate matter overrun into an adjoining blank space.

[of all.
Till one man's weakness grows the strength

Argentina: [710
Wireless, regulations of-- 93, 682, 703,

**9.8.** When matter in brackets makes more than one paragraph, start each paragraph with a bracket and place the closing bracket at end of last paragraph.

## Colon

***The colon is used—***

**9.9.** Before a final clause that extends or amplifies preceding matter. (See also rule 9.49, p. 142.)

Give up conveniences; do not demand special privileges; do not stop work: these are necessary while we are at war.
Railroading is not a variety of outdoor sport: it is service.

**9.10.** To introduce formally any matter which forms a complete sentence, question, or quotation.

The following question came up for discussion: What policy should be adopted?
He said: [if direct quotation of more than a few words follows]. (See also rule 9.49, p. 142.)
There are three factors, as follows: First, military preparation; second, industrial mobilization; and third, manpower.

**9.11.** After a salutation.

MY DEAR SIR:
*Ladies and Gentlemen:*
*To Whom It May Concern:*

**9.12.** In expressing clock time.

2:40 p.m. (use thin colon; if not available, do not use thin space on right of colon)

**9.13.** After introductory lines in lists, tables, and leaderwork, if subentries follow.

Seward Peninsula:
  Council district:
    Northern Light Mining Co.
    Wild Goose Trading Co.
  Fairhaven district: Alaska Dredging Association (single subitem runs in).
Seward Peninsula: Council district (single subitem runs in):
  Northern Light Mining Co.
  Wild Goose Trading Co.

**9.14.** In Biblical and other citations (full space after colon).

Luke 4: 3.
I Corinthians xiii: 13.
Journal of Education 3: 342–358.

**9.15.** In bibliographic references, between place of publication and name of publisher.

Congressional Directory. Washington: U.S. Government Printing Office.

**9.16.** To separate book titles and subtitles.

Financial Aid for College Students: Graduate
Germany Revisited: Education in the Federal Republic

**9.17.** In imprints before the year (en space each side of colon).

U.S. Government Printing Office
Washington : 1966

**9.18.** In proportions.

Concrete mixed 5:3:1 (use 9-unit center colons)
*but* 5–2–1 (when so in copy)

**9.19.** In double colon as ratio sign.

1:2::3:6 (use 9-unit center colon for single colons; thin colons to make double colon, with thin space each side of double colon)

## Comma

***The comma is used—***

**9.20.** To separate two words or figures that might otherwise be misunderstood.

Instead of hundreds, thousands came.
Instead of 20, 50 came.
February 10, 1929.
In 1930, 400 men were dismissed.
To John, Smith was very kind.
What the difficulty is, is not known; *but* he suggested that that committee be appointed.

**9.21.** Before a direct quotation of only a few words following an introductory phrase. (See also rule 9.10, p. 138.)

He said, "Now or never."

**9.22.** To indicate the omission of a word or words.

Then we had much; now, nothing

**9.23.** After each of a series of coordinate qualifying words.

short, swift streams; *but* short tributary streams

**9.24.** Between introductory modifying phrase and subject modified.

Beset by the enemy, they retreated.

**9.25.** Before and after *Jr.*, *Sr.*, *Esq.*, *Ph. D.*, *F.R.S.*, etc., within a sentence.

Henry Smith, Jr., chairman
Peter Johns, F.R.S., London
Washington, D.C., schools
Motorola, Inc., factory
Brown, A. H., Jr. (*not* Brown, Jr., A. H.)
*but* John Smith 2d (*or* II); Smith, John, II
Mr. Smith, Junior, also spoke (where only last name is used)

**9.26.** To set off parenthetic words, phrases, or clauses.

Mr. Jefferson, who was then Secretary of State, favored the location of the National Capital at Washington.
It must be remembered, however, that the Government had no guarantee.
It is obvious, therefore, that this office cannot function.
The atom bomb, which was developed at the Manhattan project, was first used in World War II.
Their high morale might, he suggested, have caused them to put success of the team above the reputation of the college.
The restriction is laid down in title IX, chapter 8, section 15, of the code.

*but:*

The man who fell [restrictive clause] broke his back.
The dam which gave way [restrictive clause] was poorly constructed.
He therefore gave up the search.

**9.27.** To set off words or phrases in apposition or in contrast.

Mr. Green, the lawyer, spoke for the defense.
Mr. Jones, attorney for the plaintiff, signed the petition.
Mr. Smith, not Mr. Black, was elected.
James Roosevelt, Democrat, of California.

**9.28.** After each member within a series of three or more words, phrases, letters, or figures used with *and, or,* or *nor.*

red, white, and blue
horses, mules, and cattle; *but* horses and mules and cattle
by the bolt, by the yard, or in remnants
a, b, and c
six, seven, and 10
neither snow, rain, nor heat
2 days, 3 hours, and 4 minutes (series); *but* 2 days 3 hours 4 minutes (age)

**9.29.** Before the conjunction in a compound sentence with an independent clause.

Fish, mollusks, and crustaceans were plentiful in the lakes, and turtles frequented the shores.
The boy went home alone, and his sister remained with the crowd.

**9.30.** After a noun or phrase in direct address.

Senator, will the measure be defeated?
Mr. Chairman, I will reply to the gentleman later.

**9.31.** After an interrogative clause, followed by a direct question.

You are sure, are you not? You will go, will you not?

**9.32.** Between title of person and name of organization in the absence of the words *of* or *of the.* (See also rule 9.45, p. 141.)

Chief, Division of Finance
chairman, Committee on Appropriations
colonel, 7th Cavalry
president, Yale University

**9.33.** Inside closing quotation mark. (This is not intended to change existing practice in bills and other legislative work.) (See also rules 9.130–9.131, p. 150.)

He said "four," not "five."
"Freedom is an inherent right," he insisted.
Items marked "A," "B," and "C," inclusive, were listed.

**9.34.** To separate thousands, millions, etc., in numbers of four or more digits. (See also rule 9.39, p. 141.)

4,230 50,491 1,000,000

**9.35.** After year in complete dates within sentence.

The reported dates of September 11, 1943, to June 12, 1944, were proved erroneous; *but* production for June 1950 was normal.

***The comma is omitted—***

**9.36.** Before ZIP (zone improvement plan) postal-delivery number.

Government Printing Office, Washington, D.C. 20401
Washington, D.C. 20401, is the * * *

For single lines, see rule 17.1, p. 221.

**9.37.** Between month and year in dates.

June 1938; 22d of May 1938; February and March 1938; January, February, and March 1938; January 24 A.D. 1938; 15th of June A.D. 1938; 150 B.C.; Labor Day 1966; Easter Sunday 1966; 5 January 1944 (military usage)

**9.38.** Between the name and number of an organization.

Columbia Typographical Union No. 101
General U. S. Grant Post No. 25

**9.39.** In built-up fractions, in decimals, and in serial numbers, except patent numbers.

1/2500
1.0947
page 2632
Metropolitan 5–9020 (telephone number)
1721–1727 St. Clair Avenue
Executive Order 11242
motor No. 189463
1450 kilocycles; 1100 meters (no comma unless more than four figures radio only)

**9.40.** Between superior figures or letters in footnote references.

Numerous instances may be cited.[1 2]
Data are based on October production.[a b]

**9.41.** Between two nouns one of which identifies the other.

The Children's Bureau's booklet "Infant Care" is a bestseller.

**9.42.** Before ampersand (&). (For exception, see rule 16.32, p. 219.)

Brown, Wilson & Co.
Mine, Mill & Smelter Workers

**9.43.** Before a dash.

**9.44.** In bibliographies, between name of publication and volume or similar number.

American Library Association Bulletin 34: 238, April 1940.

**9.45.** Wherever possible without danger of ambiguity.

$2 gold
$2.50 U.S. currency
$3.50 Mexican
Executive Order No. 21
General Order No. 12; *but* General Orders, No. 12
Public Law 85–1
My age is 30 years 6 months 12 days
John Lewis 2d (*or* II)
Martin of Alabama; Martin of Massachusetts; *but* Robert F. Kennedy, of New York; Kennedy of Massachusetts (duplicate names of Senators or Representatives in U.S. Congress)
Carroll of Carrollton; Henry of Navarre (places closely identified with the persons); *but* John Anstruther, of New York; President Hadley, of Yale University
James Bros. et al.; *but* James Bros., Nelson Co., et al. (last element of series)

## Dash

***The em dash is used—***

**9.46.** To mark a sudden break or abrupt change in thought. (See also rule 9.81, p. 145.)

He said—and no one contradicted him—"The battle is lost."
If the bill should pass—which God forbid!—the service will be wrecked.
The auditor—shall we call him a knave or a fool?—approved an inaccurate statement.

**9.47.** To indicate an interruption or an unfinished word or sentence. A 2-em dash will be used when the interruption is by a person other than the speaker and a 1-em dash will show self-interruption. (Note that extracts must begin with a true paragraph. Following extracts, 10 point must start as a paragraph, as example shown.)

"Such an idea can scarcely be——"
"The word 'donation'——"
"The word 'dona——' "
He said: "Give me lib——"
The bill reads "repeal," not "am——"
Q. Did you see——A. No, sir.

Mr. BROWN (reading):
The report goes on to say that—
Observe this closely—
during the fiscal year * * *.

**9.48.** Instead of commas or parentheses, if the meaning may thus be clarified.

These are shore deposits—gravel, sand, and clay—but marine sediments underlie them.

**9.49.** Before a final clause that summarizes a series of ideas. (See also rule 9.9, p. 138.)

Freedom of speech, freedom of worship, freedom from want, freedom from fear—these are the fundamentals of moral world order.

**9.50.** After an introductory phrase reading into the following lines and indicating repetition of such phrase.

I recommend—
That we accept the rules;
That we also publish them; and
That we submit them for review.

**9.51.** With a preceding question mark, in lieu of a colon.

How can you explain this?—"Fee paid, $5."

**9.52.** Sometimes, in lieu of opening quotation mark, in French, Spanish, and Italian dialog.

**9.53.** To precede a credit line or a run-in credit or signature.

Still achieving, still pursuing,
Learn to labor and to wait.
—*Longfellow.*

Every man's work shall be made manifest.—I Corinthians 3: 13.
This statement is open to question.—GERALD H. FORSYTHE.

**9.54.** After a run-in sidehead. (For examples, see rule 9.96, p. 146.)

**9.55.** To separate run-in questions and answers in testimony. (See also rule 18.27, p. 230.)

Q. Did he go?—A. No.

***The em dash is not used—***

**9.56.** At the beginning of any line of type, except as indicated in paragraphs 9.51 and 9.52.

**9.57.** Immediately after a comma, colon, or semicolon.

***The en dash is used—***

**9.58.** In a combination of figures, letters, or figures and letters. (An en dash, not a hyphen, is used, even when such terms are adjective modifiers.) (See also rule 6.44, p. 79.)

exhibit 6–A
5–20 bonds
DC–14; *but* Convair 340
Public Law 85–1, *but* Public Laws 85–1—85–20 (note em dash between two elements with en dashes)

4–H Club
LK–66–A(2)–74
$15–$20
CBS–TV network
AFL–CIO merger
*but* ACF-Brill Motors Co.

**9.59.** In the absence of the word *to* when denoting a period of time. (See also rule 11.8c, p. 170.)

1935–37 January–June Monday–Friday

***The en dash is not used—***

**9.60.** For *to* when the word *from* precedes the first of two related figures or expressions. (See also rule 11.8c, p. 170.)

From January 1 to June 30, 1951; *not* from January 1–June 30, 1951.

**9.61.** For *and* when the word *between* precedes the first of two related figures or expressions.

Between 1923 and 1929; *not* between 1923–29

## Ellipsis

**9.62.** Three asterisks, separated by en quads, are used to denote an ellipsis within a sentence, at the beginning or end of a sentence, or in two or more consecutive sentences (see also rule 9.69). If periods are used instead of asterisks, they are also separated by en quads. To achieve faithful reproduction of excerpt material, editors should indicate placement of terminal period in relation to an ellipsis at the end of a sentence.

He called * * * and left.#* * *#When he returned the * * *.

* * * called * * * and left.#* * * he returned the * * *.

He called * * * and * * *.□When he returned the * * *.

He called * * * and * * * he returned the * * *. (Two or more consecutive sentences, including intervening punctuation)

**9.63.** Three periods may be used to indicate ellipsis; four periods, when sentence is brought to a close.

He called . . . and left.#. . .#When he returned the . . . .

. . . called . . . and left.#. . . he returned the . . . .

He called . . . and . . . .□When he returned the . . . .

He called . . . and . . . he returned the . . . . (Two or more consecutive sentences, including intervening punctuation)

**9.64.** Neither asterisks nor periods are overrun alone at the end of a paragraph.

**9.65.** When both asterisks and periods for ellipsis occur in the copy and periods are not specifically requested, use asterisks throughout.

**9.66.** A line of asterisks (or periods) indicates an omission of one or more entire paragraphs. In 26½-pica or wider measure, a line of "stars" means 7 asterisks indented 2 ems at each end of the line, with the remaining space divided evenly between the asterisks. In measures less than 26½ picas, 5 stars are used. Quotation marks are not used on line of asterisks or periods in quoted matter. Where line of asterisks ends complete quotation, no closing quote is used.

* * * * * * *

. . . . . . .

**9.67.** Indented matter in 26½-pica or wider measure also requires a 7-star line to indicate the omission of one or more entire paragraphs.

**9.68.** An extra indention is added in indented matter; except where there are too many varying indentions, then all the asterisks (or periods) have the same alinement.

**9.69.** If an omission occurs in the last part of a paragraph immediately before a line of stars, three stars are used, in addition to the line of stars, to indicate such an omission.

**9.70.** If two or more sizes of type are used on a page, 10-point asterisks are indented 2 ems, 8 point and 6 point being alined with the 10-point asterisks.

**9.71.** Equalize spacing above and below a line of stars.

## Exclamation point

**9.72.** The exclamation point is used to mark surprise, incredulity, admiration, appeal, or other strong emotion, which may be expressed even in a declarative or interrogative sentence.

He acknowledged the error!
How beautiful!
"Great!" he shouted. (Note omission of comma.)
What!
Who shouted, "All aboard!" (Note omission of question mark.)

**9.73.** In direct address, either to a person or a personified object, *O* is used without an exclamation point, or other punctuation; but if strong feeling is expressed, an exclamation point is placed at the end of the statement.

O my friend, let us consider this subject impartially.
O Lord, save Thy people!

**9.74.** In exclamations without direct address or appeal, *oh* is used instead of *O*, and the exclamation point is omitted.

Oh, but the gentleman is mistaken.
Oh dear; the time is so short.

## Hyphen

***The hyphen (a punctuation mark, not an element in the spelling of words) is used—***

**9.75.** To connect the elements of certain compound words. (See "Compound Words," pp. 73–80.)

**9.76.** To indicate continuation of a word divided at end of a line. (See Word Division, supplement to STYLE MANUAL; for brief description of supplement, see p. 2. For foreign languages, see separate foreign languages.)

**9.77.** Between the letters of a spelled word.

c-e-n-t-s h-o-l-d-u-p

**9.78.** To separate elements of chemical formulas. (See rule 6.44, p. 79.)

**9.79.** If a break in six digits or over is unavoidable, divide on the comma, retain it, and use a hyphen.

## Parentheses

***Parentheses are used—***

**9.80.** To set off matter not intended to be part of the main statement or not a grammatical element of the sentence, yet important enough to be included. (For use of brackets, see rule 9.4, p. 137.)

This case (124 U.S. 329) is not relevant.
The result (see fig. 2) is most surprising.
The United States is the principal purchaser (by value) of these exports (23 percent in 1955 and 19 percent in 1956).
(Discussion off the record.)
(Pause.)
The WITNESS (interrupting). It is known * * *.
Mr. JONES (continuing). Now let us take the next item.
Mr. SMITH (presiding).
Mr. JONES (interposing).

(The matter referred to is as follows:)
The CHAIRMAN (to Mr. Smith).
The CHAIRMAN (reading):
Mr. KELLEY (to the chairman). From 15 to 25 percent.
(Objected to.)
(Mr. Smith nods.)
(Mr. Smith aside.)
(Mr. Smith makes further statement off the record.)
Mr. JONES (for Mr. Smith).
A VOICE FROM AUDIENCE. (Use of caps and small caps in hearings.)
SEVERAL VOICES. (Use of caps and small caps in hearings.)

**9.81.** To enclose a parenthetic clause where the interruption is too great to be indicated by commas. (See also rule 9.46, p. 141.)

You can find it neither in French dictionaries (at any rate, not in Littré) nor in English.

**9.82.** To enclose an explanatory word not part of the statement.

the Erie (Pa.) News; *but* the News of Erie, Pa.
Portland (Oreg.) Chamber of Commerce; *but* Washington, D.C., schools.

**9.83.** To enclose letters or numbers designating items in a series, either at beginning of paragraphs or within a paragraph.

The order of delivery will be: (*a*) Food, (*b*) clothing, and (*c*) tents and other housing equipment.
You will observe that the sword is (1) old fashioned, (2) still sharp, and (3) unusually light for its size.
Paragraph 7(B)(1)(*a*) will be found on page 6. (Note parentheses closed up (see rule 2.9, p. 8).)

**9.84.** To enclose a figure inserted to confirm a statement given in words if double form is specifically requested. (See also rule 11.18, p. 173.)

This contract shall be completed in sixty (60) days.

**9.85.** A reference in parentheses at the end of a sentence is placed before the period, unless it is a complete sentence in itself.

The specimen exhibits both phases (pl. 14, *A*, *B*).
The individual cavities show great variation. (See pl. 4.)

**9.86.** If a sentence contains more than one parenthetic reference, the one at the end is placed before the period.

This sandstone (see pl. 6) occurs in every county of the State (see pl. 1).

**9.87.** When a figure is followed by a letter in parentheses, no space is used between the figure and the opening parenthesis; but if the letter is not in parentheses and the figure is repeated with each letter, the letter is closed up with the figure.

15(*a*). Classes, grades, and sizes.
15*a*. Classes, grades, and sizes.

**9.88.** If both a figure and a letter in parentheses are used before each paragraph, a period and an en space are used after the closing parenthesis; if the figure is not repeated before each letter in parentheses but is used only before the first, the period is placed after the figure.

15(*a*). When the figure is used before the letter in each paragraph—
15(*b*). The period is placed after the closing parenthesis.
15. (*a*) When the figure is used before letter in first paragraph but not repeated with subsequent letters—
(*b*) The period is used after the figure only.

**9.89.** Note position of period relative to closing parenthesis:

The vending stand sells a variety of items (sandwiches, beverages, cakes, etc.).
The vending stand sells a variety of items (sandwiches, beverages, cakes, etc. (sometimes ice cream)).
The vending stand sells a variety of items. (These include sandwiches, beverages, cakes, etc. (6).)

**9.90.** To enclose bylines in congressional work.

(By Sylvia Porter, staff writer)

**9.91.** When matter in parentheses makes more than one paragraph, start each paragraph with a parenthesis and place the closing parenthesis at end of last paragraph.

## Period

***The period is used—***

**9.92.** After a declarative sentence that is not exclamatory or after an imperative sentence.

Stars are suns.
He was employed by Sampson & Co.
Do not be late.
On with the dance.

**9.93.** After an indirect question or after a question intended as a suggestion and not requiring an answer.

Tell me how he did it.
May we hear from you.
May we ask prompt payment.

**9.94.** In place of parentheses after a letter or number denoting a series.

*a.* Bread well baked.
*b.* Meat cooked rare.
*c.* Cubed apples stewed.

1. Punctuate freely.
2. Compound sparingly.
3. Index thoroughly.

**9.95.** Sometimes to indicate ellipsis. (See rule 9.63, p. 143.)

**9.96.** After a run-in sidehead.

*Conditional subjunctive.*—The conditional subjunctive is required for all unreal and doubtful conditions.

**2. Peacetime preparation.**—*a.* The Chairman of the National Security Resources Board, etc.

2. *Peacetime preparation—Industrial mobilization plans.*—The Chairman of the National Security Resources Board, etc.

**2. Peacetime preparation.**—*Industrial mobilization.*—The Chairman of the National Security Resources Board, etc.

62. *Determination of types.*—*a. Statement of characteristics.*—Before types of equipment, etc.

**Steps in planning for procurement.**—(1) *Determination of needs.*—To plan for the procurement of such arms, etc.

62. *Determination of types.*—(*a*) *Statement of characteristics.*—Before types of, etc.

**DETERMINATION OF TYPES.—Statement of characteristics.**—Before types of, etc.

NOTE.—The source material was furnished.
*but* Source: U.S. Department of Commerce, Bureau of the Census.

**9.97.** Paragraphs and subparagraphs may be arranged according to the scheme below. The sequence is not fixed, and variations, in addition to the use of center and side heads or indented paragraphs, may be adopted, depending on the number of parts.

I. (Roman numeral)
A.
1.
*a.*
(1)
(*a*)
(i) (lowercase Roman numeral)
(*aa*)

**9.98.** To separate integers from decimals in a single expression.

3.75 percent $3.50 1.25 meters

**9.99.** In continental European languages, to indicate thousands.

1.317 72.190.175

**9.100.** After abbreviations, unless otherwise specified. (See "Abbreviations," p. 153.)

| gal. | NE. | m. (meter) |
|---|---|---|
| qt. | N.Y. | kc. (kilocycle) |

**9.101.** After legends and explanatory matter beneath illustrations. However, legends without descriptive language do not require periods.

FIGURE 1.—Schematic drawing.
FIGURE 1.—Continued.
*but* FIGURE 1 (no period)

**9.102.** Rarely, to indicate multiplication. (The multiplication sign is preferable for this purpose.)

$a{\cdot}b$ $(a\times b)$

**9.103.** After *Article 1*, *Section 1*, etc., at beginning of paragraphs. A period and en space are used after such terms.

***The period is omitted—***

**9.104.** After—

Lines in title pages.
Center, side, and running heads.
Continued lines.
Boxheads of tables.
Scientific, chemical, or other symbols.

This rule does not apply to abbreviation periods.

**9.105.** After a quotation mark that is preceded by a period. (See also rule 9.131, p. 150.)

He said, "Now or never."

**9.106.** After letters used as names without specific designation.

A said to B that all is well.
Mr. A told Mr. B that the case was closed.
*but* Mr. A. (for Mr. Andrews). I do not want to go.
Mr. K. (for Mr. King). The meeting is adjourned.

**9.107.** After a middle initial which is merely a letter and not an abbreviation of a name.

Daniel D Tompkins Ross T McIntire

**9.108.** After a short name which is not an abbreviation of the longer form. (See also rule 10.23, p. 156.)

Alex Ed Sam

**9.109.** After Roman numerals used as ordinals.

George V

**9.110.** After words and incomplete statements listed in columns. Full-measure matter is not to be regarded as a column.

**9.111.** After explanatory matter set in 6 point under leaders or rules.

------------------ -------------------- ---------------------
(Name) (Address) (Position)

**9.112.** Immediately before leaders, even if an abbreviation precedes the leaders.

## Question mark

***The question mark is used—***

**9.113.** To indicate a direct query, even if not in the form of a question.

Did he do it?
He did what?
Can the money be raised? is the question.
Who asked, "Why?" (Note single question mark)
"Did you hurt yourself, my son?" she asked.

**9.114.** To express more than one query in the same sentence.

Can he do it? or you? or anyone?

**9.115.** To express doubt.

He said the boy was 8(?) feet tall. (No space before question mark)
The statue(?) was on the statute books.

## Quotation marks

***Quotation marks are used—***

**9.116.** To enclose direct quotations. (Each part of an interrupted quotation begins and ends with quotation marks.)

The answer is "No."
He said, "John said 'No.' "
"John," said Henry, "why do you go?"

**9.117.** To enclose any matter following the terms *entitled, the word, the term, marked, designated, classified, named, endorsed,* or *signed;* but are not used to enclose expressions following the terms *known as, called, so-called,* etc., unless such expressions are misnomers or slang.

Congress passed the act entitled "An act * * *."
After the word "treaty," insert a comma.
Of what does the item "Miscellaneous debts" consist?
The column "Imports from foreign countries" was not * * *.
The document will be marked "Exhibit No. 21"; *but* The document may be made exhibit No. 2.
The check was endorsed "John Adamson."
It was signed "John."
Beryllium is known as glucinium in some European countries.
It was called profit and loss.
The so-called investigating body.

**9.118.** To enclose titles of addresses, articles, books, captions, chapter and part headings, editorials, essays, headings, headlines, motion pictures and plays (including TV and radio programs), papers, short poems, reports, songs, subheadings, subjects, and themes. All principal words are to be capitalized. (See also rule 3.52, p. 31.)

An address on "Uranium-235 in the Atomic Age"
The article "Germany Revisited," appeared in the last issue

"The Conquest of Mexico," a published work (book)
Under the caption "Long-Term Treasurys Rise"
The subject was discussed in "Courtwork" (chapter heading)
It will be found in "Part XI: Early Thought"
The editorial "Haphazard Budgeting"
"Compensation," by Emerson (essay)
"United States To Appoint Representative to U.N." (heading or headline)
In "Search for Paradise" (motion picture); "South Pacific" (play)
A paper on "Constant-Pressure Combustion" was read
"O Captain! My Captain!" (short poem)
The report "Atomic Energy: What It Means to the Nation"; *but* annual report of the Public Printer
This was followed by the singing of "The Star-Spangled Banner"
Under the subhead, "Sixty Days of Turmoil," will be found * * *
The subject (or theme) of the conference is "Peaceful Uses of Atomic Energy"
*also* Account 5, "Management fees."
Under the heading "Management and operation."
Under the appropriation "Building of ships, Navy."

**9.119.** If poetry is quoted, each stanza should start with quotation marks, but only the last stanza should end with them. The lines of the poem should range on the left, those that rhyme taking the same indention, and the quotation marks should be cleared. Poems are centered on the longest line; overs 3 ems; 2 leads between stanzas.

"Rest is not quitting
The busy career;
Rest is the fitting
Of self to one's sphere.

"'Tis the brook's motion,
Clear without strife,
Fleeing to ocean
After its life."
—*John Sullivan Dwight.*

**9.120.** At the beginning of each paragraph of a quotation, but at the end of the last paragraph only.

**9.121.** To enclose a letter or other communication, which bears both date and signature, within a letter. (See rule 9.126.)

**9.122.** To give greater emphasis to a word or a phrase. (For better typographical appearance and legibility, such use of quotation marks should be kept to a minimum.)

**9.123.** To enclose misnomers, slang expressions, sobriquets, or ordinary words used in an arbitrary way. (See also rule 6.26, p. 76.)

He voted for the "lameduck" amendment.
His report was "bunk."
It was a "gentlemen's agreement."
The "invisible government" is responsible.
George Herman "Babe" Ruth.

**9.124.** Quotation marks will not be borne off from adjacent characters except when they precede a fraction or an apostrophe or precede or follow a superior figure or letter, in which case a thin space will be used. A thin space will also be used to separate double and single quotation marks.

***Quotation marks are not used—***

**9.125.** To enclose names of newspapers or magazines.

**9.126.** To enclose complete letters having date and signature.

**9.127.** To enclose extracts that are indented or set in smaller type, or solid extracts in leaded matter; but indented matter in text that is already quoted carries quotation marks.

**9.128.** In indirect quotations.

Tell her yes.
He could not say no.

**9.129.** Before a display initial which begins a quoted paragraph.

**9.130.** The comma and the final period will be placed inside the quotation marks. Other punctuation marks should be placed inside the quotation marks only if they are a part of the matter quoted. (See rule 9.33, p. 140.)

Ruth said, "I think so."
"The President," he said, "will veto the bill."
The trainman shouted, "All aboard!"
Who asked, "Why?"
The President suggests that "an early occasion be sought * * *."
Why call it a "gentlemen's agreement"?

**9.131.** In congressional and certain other classes of work showing amendments, and in courtwork with quoted language, punctuation marks are printed after the quotation marks when not a part of the quoted matter.

Insert the words "growth", "production", and "manufacture".
To be inserted immediately after the words "cadets, U.S. Coast Guard;".
Change "February 1, 1951", to "June 30, 1951".
"Insert in lieu thereof 'July 1, 1953,'."

**9.132.** When occurring together, quotation marks should precede the footnote reference number.

The commissioner claimed that the award was "unjustified."[1]
His exact words were: "The facts in the case prove otherwise."[2]

**9.133.** Quotation marks should be limited, if possible, to three sets (double, single, double).

"The question in the report is, 'Can a person who obtains his certificate of naturalization by fraud be considered a "bona fide" citizen of the United States?' "

## Semicolon

***The semicolon is used—***

**9.134.** To separate clauses containing commas. (See also rule 9.137, p. 151.)

Donald A. Peters, president of the First National Bank, was also a director of New York Central; Harvey D. Jones was a director of Oregon Steel Co. and New York Central; Thomas W. Harrison, chairman of the board of McBride & Co., was also on the board of Oregon Steel Co.
Reptiles, amphibians, and predatory mammals swallow their prey whole or in large pieces, bones included; waterfowl habitually take shellfish entire; and gallinaceous birds are provided with gizzards that grind up the hardest seeds.
Yes, sir; he did see it.
No, sir; I do not recall.

**9.135.** To separate statements that are too closely related in meaning to be written as separate sentences, and also statements of contrast.

Yes; that is right.
No; we received one-third.
It is true in peace; it is true in war.
War is destructive; peace, constructive.

**9.136.** To set off explanatory abbreviations or words which summarize or explain preceding matter.

The industry is related to groups that produce finished goods; i.e., electrical machinery and transportation equipment.

There were involved three metal producers; namely, Jones & Laughlin, Armco, and Kennecott.

**9.137.** The semicolon is to be avoided where a comma will suffice.

Regional offices are located in New York, N.Y., Chicago, Ill., and Dallas, Tex.

## Single punctuation

**9.138.** Single punctuation is used wherever possible without ambiguity.

124 U.S. 321 (no comma)
SIR: (no dash)
Joseph replied, "It is a worthwhile effort." (no outside period)

## Type

**9.139.** Parentheses, brackets, and superior reference figures are always set in roman, not in italic. All other punctuation marks match the type of the words which they adjoin. A lightface dash is used after a run-in boldface sidehead followed by lightface matter. In boldface matter, punctuation, parentheses, brackets, dashes, shilling marks, and fractions are all set in boldface, if available. (See rule 12.16, p. 176.)

# 11. NUMERALS

(See also Tabular Work; Leaderwork)

**11.1.** Most rules for the use of numerals are based on the general principle that the reader comprehends numerals more readily than numerical word expressions, particularly in technical, scientific, or statistical matter. However, for special reasons numbers are spelled out in indicated instances.

**11.2.** The following rules cover the most common conditions that require a choice between the use of numerals and words. Some of them, however, are based on typographic appearance rather than on the general principle stated above.

**11.3.** Arabic numerals are generally preferable to Roman numerals.

## NUMBERS EXPRESSED IN FIGURES

**11.4.** A figure is used for a number of *10* or more with the exception of the first word of the sentence. Numbers under *10* are to be spelled, except for time, measurement, and money. (See also rules 11.8, p. 170; 11.24, p. 173.)

| | | |
|---|---|---|
| 50 ballots | 24 horses | about 40 men |
| 10 guns | nearly 10 miles | 10 times as large |

Each of 15 major commodities (nine metal and six nonmetal) was in supply.

Petroleum came from 16 fields, of which eight were discovered in 1956.

That man has three suits, two pairs of shoes, and 12 pairs of socks.

Of the 13 engine producers, six were farm equipment manufacturers, six were principally engaged in the production of other types of machinery, and one was not classified in the machinery industry.

There were three six-room houses, five four-room houses, and three two-room cottages, and they were built by 20 men. (See rule 11.22, p. 173.)

There were three six-room houses, five four-room houses, and three two-room cottages, and they were built by nine men.

Only four companies in the metals group appear on the list, whereas the 1947 census shows at least 4,400 establishments.

*but* If two columns of sums of money add or subtract one into the other and one carries points and ciphers, the other should also carry points and ciphers.

At the hearing, only one Senator and one Congressman testified.

There are four or five things which can be done.

**11.5.** A unit of measurement, time, or money (as defined in rule 11.8, p. 170) is always expressed in figures.

Each of the five girls earned 75 cents an hour.

Each of the 15 girls earned 75 cents an hour.

A team of four men ran the 1-mile relay in 3 minutes 20 seconds.

This usually requires from two to five washes and a total time of 2 to 4 hours.

This usually requires nine to 12 washes and a total time of 2 to 4 hours.

The contractor, one engineer, and one surveyor inspected the 1-mile road.

*but* There were two six-room houses, three four-room houses, and four two-room cottages, and they were built by nine men in thirty 5-day weeks. (See rule 11.22, p. 173.)

789-445°—67——12

### 11.6. Figures are used for serial numbers.

Bulletin 725
Document 71
pages 352–357
lines 5 and 6
paragraph 1
chapter 2

290 U.S. 325
Genesis 39: 20 (full space after colon)
Metropolitan 5–9020 (telephone number)
the year 1931
1721–1727 St. Clair Avenue
*but* Letters Patent No. 2,189,463

### 11.7. A colon preceding figures does not affect their use.

The result was as follows: 12 voted yea, four dissented.
The result was as follows: nine voted yea, four dissented.

## Measurement and time

### 11.8. Units of measurement and time are expressed in figures.

**a.** Age:
6 years old
52 years 10 months 6 days
a 3-year-old

**b.** Clock time (see also Time):
4:30 p.m. (use thin colon)
10 o'clock *or* 10 p.m. (*not* 10 o'clock p.m.; 2 p.m. in the afternoon; 10:00 p.m.); 12 m. (noon); 12 p.m. (midnight)
half past 4
$4^h30^m$ *or* $4.5^h$, in scientific work, if so written in copy
0025, 2359 (astronomical and military time)

**c.** Dates:
June 1935; June 29, 1935 (*not* June, 1935, *nor* June 29th, 1935)
March 6 to April 15, 1935 (*not* March 6, 1935, to April 15, 1935)
May, June, and July 1935 (*but* June and July 1935)
15 April 1951 (military)
the 2d (*or* 3d) instant
4th of July (*but* Fourth of July, meaning the holiday)
the 1st [day] of the month (*but* the last of April or the first of May, not referring to specific days)

In referring to a fiscal year, consecutive years, or a continuous period of 2 years or more, when contracted, the forms 1906–38, 1931–32, 1801–2, 1875–79 are used (*but* 1895–1914, 1900–1901); for two or more separate years not representing a continuous period, a comma is used instead of a dash (1875, 1879); if the word *from* precedes the year or the word *inclusive* follows it, the second year is not shortened and the word *to* is used in lieu of the dash (from 1933 to 1936; 1935 to 1936, inclusive).

In dates, *A.D.* precedes the year (A.D. 937); *B.C.* follows the year (254 B.C.).

**d.** Decimals: In text a cipher should be supplied before a decimal point if there is no unit, and ciphers should be omitted after a decimal point unless they indicate exact measurement.
0.25 inch; 1.25 inches
silver 0.900 fine
specific gravity 0.9547
gage height 10.0 feet
*but* .30 caliber (meaning 0.30 inch, bore of small arms); 30 calibers (length)

**e.** Degrees, etc. (spaces omitted):
longitude 77°04′06″ E.
latitude 49°26′14″ N.
35°30′; 35°30′ N.
a polariscopic test of 85°
45.5° to 49.5° below zero
an angle of 57°
strike N. 16° E.
dip 47° W. *or* 47° N. 31° W.
gravity 16.6° B.
25.5′ (preferred); *also* 25′.5 *or* 25′.5, as in copy
*but* two degrees of justice; 12 degrees of freedom
32d degree Mason

**f.** Market quotations:
4½-percent bonds
Treasury bonds sell at 95
Metropolitan Railroad, 109
gold is 109
wheat at 2.30
sugar, .03; *not* 0.03

**g.** Mathematical expressions:
multiplied by 3
divided by 6

**h.** Measurements:

7 meters
about 10 yards
8 by 12 inches
8- by 12-inch page
2 feet by 1 foot 8 inches by 1 foot 3 inches
1½ miles
6 acres
9 bushels
1 gallon
3 ems
20/20 (vision)
60μ
2,500 horsepower
15 cubic yards
6-pounder
80 foot-pounds
10s (for yarns and threads)
*but* tenpenny nail; fourfold; three-ply; five votes; six bales (see also rule 11.23)

**i.** Money:

$3.65; $0.75; 75 cents; 0.5 cent
$3 (*not* $3.00) per 200 pounds
75 cents apiece
Rs32,25,644 (Indian rupees)
2.5 francs *or* fr2.5
£2 4s. 6d.
T£175
65 yen
₱265

**j.** Percentage:

12 percent; 25.5 percent; 0.5 percent (*or* one-half of 1 percent)
3.65 bonds; 3.65s; 5–20 bonds; 5–20s; 4½s; 3s (see also rule 5.28, p. 71)
50–50 (colloquial expression)
5 percentage points

**k.** Proportion:

1 to 4
1:62,500 (equal space each side of colon)
1–3–5

**l.** Time (see also Clock time):

6 hours 8 minutes 20 seconds
10 years 3 months 29 days
8 days
7 minutes
1 month
*but* four centuries; three decades; three quarters (9 months)
statistics of any one year
in a year or two
four afternoons

**m.** Unit modifiers:

5-day week
8-year-old wine
8-hour day
10-foot pole
½-inch pipe
5-foot-wide entrance
10-million-peso loan
a 5-percent increase
20th-century progress
*but* two-story house
five-man board
$20 million airfield

**n.** Game scores:

1 up (golf)
3 to 2 (baseball)
7 to 6 (football), etc.

## Ordinal numbers

**11.9.** Except as indicated in rule 11.19, p. 173, and also for day preceding month, figures are used in text and footnotes to text for serial ordinal numbers beginning with *10th*. In tables, leaderwork, footnotes to tables and leaderwork, and in sidenotes, figures are used at all times. Military units are expressed in figures at all times when not the beginning of sentence, except *Corps*. (For ordinals in addresses, see rule 11.11, p. 172.)

29th of May, *but* May 29
First Congress; 82d Congress
ninth century; 20th century
Second Congressional District; 20th Congressional District
seventh region; 17th region
eighth parallel; 38th parallel
fifth ward; 12th ward
ninth birthday; 66th birthday
1st Army
2d Infantry Division
323d Fighter Wing
77th Regiment
9th Naval District
7th Fleet
7th Air Force
7th Task Force

*but* XII Corps (Army usage)
Court of Appeals for the Tenth Circuit
Seventeenth Decennial Census (title)

**11.10.** Ordinals and numerals appearing in a sentence are treated according to the separate rules dealing with ordinals and numerals standing alone or in a group. (See rules 11.4, p. 169; 11.9, p. 171; 11.24, p. 173.)

> The fourth group contained three items.
> The fourth group contained 12 items.
> The eighth and 10th groups contained three and four items, respectively.
> The eighth and ninth groups contained nine and 12 items, respectively.

**11.11.** Beginning with *10th*, figures are used in text matter for numbered streets, avenues, etc., but in tables, leaderwork, footnotes, and sidenotes, figures are used at all times, and *street*, *avenue*, etc., are abbreviated. (See also rule 10.16, p. 155.)

> First Street NW.; *also* in parentheses: (Fifth Street) (13th Street); 810 West 12th Street; North First Street; 1021 121st Street; 2031 18th Street North; 711 Fifth Avenue; 518 10th Avenue; 51–35 61st Avenue

## Fractions

(For spelled-out fractions, see rule 11.28, p. 174.)

**11.12.** Piece and em fractions (¼, ⅓, ¾, ⅜, ⅝, ⅞, 1/2954) are used in text, but the shilling mark with full-sized figures (1/4, 1/2954) may be used if specially requested. A comma should not be used in any part of a built-up fraction of four or more digits or in decimals.

**11.13.** Fractions are used in a unit modifier.

> ½-inch pipe; *not* one-half-inch pipe    ¼-mile run    ⅞-point rise

## Punctuation

**11.14.** The comma is used in a number containing four or more digits, except in serial numbers, common and decimal fractions, astronomical and military time, and kilocycles and meters of not more than four figures pertaining to radio.

## Chemical formulas

**11.15.** In chemical formulas full-sized figures are used before the symbol or group of symbols to which they relate, and inferior figures are used after the symbol. (See also rules 6.44, p. 79; 13.17, p. 179.)

$$6PbS.(Ag,Cu)_2S.2As_2S_3O_4$$

### NUMBERS SPELLED OUT

**11.16.** Numerals are spelled out at the beginning of a sentence or head. Rephrase a sentence or head to avoid beginning with figures.

> Five years ago * * *; *not* 5 years ago * * *
> Five hundred and fifty men are employed * * *; *not* 550 men are employed * * *
> "Five-Year Plan Announced"; *not* "5-Year Plan Announced" (head)
> Although 1965 may seem far off, it * * *; *not* 1965 may seem far off, it * * *
> Government employees numbering 207,843 * * *; *not* 207,843 Government employees * * *
> Benefits amounting to $69,603,566 * * *; *not* $69,603,566 worth of benefits * * *

**11.17.** In testimony, hearings, transcripts, and Q. and A. matter, figures are used immediately following Q. and A. or name of interrogator or witness for years (e.g., 1958), sums of money, decimals, street numbers, and for numerical expressions beginning with *101*.

> Mr. SMITH. 1957 was a good year.
> Mr. JONES. $1 per share was the return. Two dollars in 1956 was the alltime high. Nineteen hundred and seventy-eight may be another story.
> Mr. JONES. 92 cents.

Publication Modern Power    Editor RJB    Art No. 1969    Sheet No. 1 of 8

6 12 18 24 30 36 42 48 54 60 66 72 78 84

1 In latest expansion of
2 its Samuel Smith plant... — 14 point Futura Demibold
3
4 Southwest City improves
5 on its low-cost tradition — 24 point Futura Bold
6 By Thomas H. Richards, Senior Project Engineer — 10 point Vogue Bold
7 Superior Engineering Co., Inc., New York
8 Text 9 on 10 Times Roman
9 13½ picas
10 Southwest City has done it again. Unit No. 4, the latest at its Samuel
11 Smith Power Station, which came on the line in the spring of 1968, was installed at
12 a cost overall of $75.00 per kw.
13 This is substantially lower than the cost of any of the company's previous
14 units: the "record" was $89.00 in 1961 for Samuel Smith No. 2, a 104-mw unit.
15 But the larger size of Smith No. 4, 175 mw, does not, in itself, justify such a
16 low cost. Other ~~well-built~~ (stet) power plants have ~~reported~~ costs well above $100
17 for units of that size. Generally, it isn't until the size of the ~~steam-electric~~
18 ~~generating~~ unit reaches three hundred fifty mw that such low costs are expected.
19 Three conditions must be satisfied to get the benefits of costs as low as
20 those which the Utility Board of Southwest City has consistently been able to
21 obtain in the construction of its (3) power plants (Mesquite Creek, Coolidge
22 Street and Samuel Smith). There must be a competent consulting engineer, an ex-
23 perienced constructor and, even more imperatively, a cooperative client. No-
24 thing increases the cost of a project as much as divided responsibility between
25 the client and its consulting engineering staff. Successive demands for changes

51-50700

|  | |  |
|---|---|---|
| ≡ | below a letter | capital letter |
| ═ | below letters | small capitals |
| — | below letters | italics |
| ⊏ | to left of typing | move to this point (no indent) |
| / | through a capital | make it lower case |
| \| | between letters | separate one sapce |
| \| | through a letter | delete letter, leave sapce |
| ⁐ | between letters | close up space |
| | | delete letter, close space |
| ∧ | (caret) | insert letter or word(s) |
| | | transpose words |
| ...kw. This... | | run together (no paragraph) |
| ∿ | | transpose characters |
| | | apostrophe |
| | | quotation mark |
| = | (equal sign) | (to printer) hyphen |
| ¶ | | start paragraph |
| ⊢⊣ | | dash |
| well-built (stet) | | (stet) let it stand |
| reported | (dots under) | alternative for "stet" |
| | | delete |
| steam-electric generating | | delete (do not encircle material to be deleted) |
| three hundred fifty (circled) | | change to numerals (abbreviate) |
| These... | (wavy underline) | set boldface type |
| (3) | | change to word for numeral (spell out) |

# APPENDIX B: COPYREADING PRACTICE

Closely related to style is copyreading practice. Familiarity with standard symbols for editing and correcting typed material will save many minutes and misunderstandings. These copyreading symbols should be known and used by both the typist and the person who corrects the copy. Editors, printers, and compositors can be counted upon to know them.

The example on the next page presents most of the standard marks; explanatory notes accompany the example.

Mr. SMITH. 12.8 people.
Mr. JONES. 1240 Pennsylvania Avenue.
Mr. SMITH. Ninety-eight persons.
Q. 101 years? *But* Q. One hundred years?
A. 200 years.
Mr. SMITH. Ten-year average would be how much?

**11.18.** A spelled-out number should not be repeated in figures, except in legal documents. In such instances these forms will be observed:

five (5) dollars, *not* five dollars (5)
ten dollars ($10), *not* ten ($10) dollars

**11.19.** Numbers mentioned in connection with serious and dignified subjects such as Executive orders, legal proclamations, and in formal writing are spelled out.

the Thirteen Original States
in the year nineteen hundred and forty-four
the Seventy-eighth Congress
millions for defense but not one cent for tribute
threescore years and ten

**11.20.** Numbers expressing time, money, or measurement separated from their unit descriptions by more than two words are spelled out if under *10*.

two and more separate years
whether five or any number of years

*but* 5 successive years
4 calendar years
6 hard-earned dollars
5 up to 10 dollars

**11.21.** Numbers larger than 1,000, if spelled out, should be in the following form:

two thousand and twenty
one thousand eight hundred and fifty
one hundred and fifty-two thousand three hundred and five
eighteen hundred and fifty (serial number)

**11.22.** Numbers of less than *100* preceding a compound modifier containing a figure are spelled out.

two ¾-inch boards
twelve 6-inch guns

three four-room houses
*but* 120 8-inch boards

**11.23.** Indefinite expressions are spelled out.

the seventies; the early seventies; *but* the early 1870's *or* 1870's
a thousand and one reasons
between two and three hundred horses [1]
midthirties
in the eighties, *not* the '80's *nor* 80's

twelvefold; fortyfold; hundredfold, twentyfold to thirtyfold
*but* 1 to 3 million
mid-1951
40-odd people; nine-odd people
40-plus people
100-odd people
3½-fold; 250-fold; 2.5-fold; 41-fold

The words *nearly, about, around, approximately,* etc., do not constitute indefinite expressions.

**11.24.** Except as indicated in rule 11.8 (p. 170), a number less than *10* is spelled out within a sentence. (See rule 11.4, p. 169.)

six horses
five wells
eight times as large

*but* 3½ cans
2½ times *or* 2.5 times

**11.25.** For typographic appearance and easy grasp of large numbers beginning with *million,* the word *million* or *billion* is used.

[1] Better: Between 200 and 300 horses.

The following are guides to treatment of figures as submitted in copy. If copy reads—

$12,000,000, *change to* $12 million
2,750,000,000 dollars, *change to* $2,750 million
2.7 million dollars, *change to* $2.7 million
2⅜ million dollars, *change to* $2⅜ million
two and a half million dollars, *do not change to* $2½ million.
two and one-half million dollars, *change to* $2½ million
*but* $2,700,000, *do not change to* $2.7 million
*also* $10 to $20 million; 10 or 20 million; between 10 and 20 million; $10 million or $20 million; if in copy, follow
4 millions of assets
amounting to 4 millions
$1,270,000
$1,270,200,000
$2¾ billion; $2.75 billion; $2,750 million
$500,000 to $1 million
300,000; *not* 300 thousand
$½ billion to $1¼ billion (note full figure with second fraction); $1¼ to $1½ billion.
three-quarters of a billion dollars
5 or 10 billion dollars' worth (see rule 5.31, p. 71)

**11.26.** Related numbers close together at the beginning of a sentence are treated alike.

Fifty or sixty miles away is snowclad Mount McKinley.

**11.27.** Round numbers are spelled out.

a hundred cows
a thousand dollars
a million and a half
two thousand million dollars
less than a million dollars

**11.28.** Fractions standing alone, or if followed by *of a* or *of an*, are generally spelled out. (See also rule 11.13, p. 172.)

three-fourths of an inch; *not* ¾ inch *nor* ¾ of an inch
one-half inch
one-half of a farm; *not* ½ of a farm
one-fourth inch
*or*, if copy so reads:
three-quarters of an inch
half an inch
a quarter of an inch
one-tenth
one-hundredth
two one-hundredths
one-thousandth
five one-thousandths
thirty-five one-thousandths
*but* ½ to 1¾ pages
½-inch pipe
½-inch-diameter pipe
3½ cans; 2½ times

## ROMAN NUMERALS

**11.29.** A repeated letter repeats its value; a letter placed after one of greater value adds to it; a letter placed before one of greater value subtracts from it; a dashline over a letter denotes multiplied by 1,000.

| | | | | | | | |
|---|---|---|---|---|---|---|---|
| I | 1 | XXIX | 29 | LXXV | 75 | DC | 600 |
| II | 2 | XXX | 30 | LXXIX | 79 | DCC | 700 |
| III | 3 | XXXV | 35 | LXXX | 80 | DCCC | 800 |
| IV | 4 | XXXIX | 39 | LXXXV | 85 | CM | 900 |
| V | 5 | XL | 40 | LXXXIX | 89 | M | 1,000 |
| VI | 6 | XLV | 45 | XC | 90 | MD | 1,500 |
| VII | 7 | XLIX | 49 | XCV | 95 | MM | 2,000 |
| VIII | 8 | L | 50 | XCIX | 99 | MMM | 3,000 |
| IX | 9 | LV | 55 | C | 100 | MMMM or $M\overline{V}$ | 4,000 |
| X | 10 | LIX | 59 | CL | 150 | $\overline{V}$ | 5,000 |
| XV | 15 | LX | 60 | CC | 200 | $\overline{M}$ | 1,000,000 |
| XIX | 19 | LXV | 65 | CCC | 300 | | |
| XX | 20 | LXIX | 69 | CD | 400 | | |
| XXV | 25 | LXX | 70 | D | 500 | | |

*Dates*

| | | | | | |
|---|---|---|---|---|---|
| MDC | 1600 | MCMX | 1910 | MCML | 1950 |
| MDCC | 1700 | MCMXX | 1920 | MCMLX | 1960 |
| MDCCC | 1800 | MCMXXX | 1930 | | |
| MCM or MDCCCC | 1900 | MCMXL | 1940 | | |

# APPENDIX C: TECHNICAL AND BUSINESS MAGAZINES THAT PUBLISH CONTRIBUTED ARTICLES

Most of the publications listed have nationwide United States circulation over 10,000.

*A.I.A. Journal*
1735 New York Ave. N.W.
Washington, D.C. 20006

*Journal of Accountancy*
666 Fifth Ave.
New York, N.Y. 10019

*Actual Specifying Engineer*
1801 Prairie Ave.
Chicago, Ill. 60616

*Adhesives Age*
101 W. 31st St.
New York, N.Y. 10001

*Advertising & Sales Promotion*
740 Rush St.
Chicago, Ill. 60611

*Air Conditioning, Heating & Refrigeration News*
Box 6000
Birmingham, Mich. 48012

*Air Conditioning, Heating and Ventilating*
200 Madison Ave.
New York, N.Y. 10016

*Journal of the Air Pollution Control Association*
4400 Fifth Ave.
Pittsburgh, Pa. 15213

*Airline Management & Marketing*
One Park Ave.
New York, N.Y. 10016

*Airport Services Management*
731 Hennepin Ave.
Minneapolis, Minn. 55403

*American Ceramic Society Bulletin*
4055 High St.
Columbus, Ohio 43214

*American City Magazine*
Berkshire Commons
Pittsfield, Mass. 01201

*American Machinist*
330 W. 42nd St.
New York, N.Y. 10036

*American School & University*
Berkshire Commons
Pittsfield, Mass. 01201

*Appliance Manufacturer*
5 S. Wabash Ave.
Chicago, Ill. 60603

*Architectural and Engineering News*
56th and Chestnut Sts.
Philadelphia, Pa. 19139

*Architectural Forum*
111 W. 57th St.
New York, N.Y. 10019

*Architectural Record*
330 W. 42nd St.
New York, N.Y. 10036

*ASHRAE Journal*
345 E. 47th St.
New York, N.Y. 10017

*Asphalt*
Asphalt Institute Bldg.
College Park, Md. 20740

*Assembly Engineering*
Hitchcock Bldg.
Wheaton, Ill. 60187

*Astronautics and Aeronautics*
1290 Avenue of the Americas
New York, N.Y. 10019

*Automatic Machining*
65 Broad St.
Rochester, N.Y. 14614

*Automation*
Penton Bldg.
Cleveland, Ohio 44113

*Automotive Age*
6347 Fountain Ave.
Hollywood, Calif. 90028

*Automotive Industries*
Chestnut & 56th Sts.
Philadelphia, Pa. 19139

*Aviation Week & Space Technology*
330 W. 42nd St.
New York, N.Y. 10036

*Barron's National Business & Financial Weekly*
30 Broad St.
New York, N.Y. 10004

*Better Roads*
333 N. Michigan Ave.
Chicago, Ill. 60601

*Bio Science*
3900 Wisconsin Ave. N. W.
Washington, D.C. 20016

*Blast Furnace & Steel Plant*
624 Grant Bldg.
Pittsburgh, Pa. 15231

*Boxboard Containers*
300 W. Adams St.
Chicago, Ill. 60606

*Building Construction*
5 S. Wabash Ave.
Chicago, Ill. 60603

*Bulletin of the Atomic Scientists*
935 E. 60th St.
Chicago, Ill. 60637

*Business Automation*
288 Park Ave. W.
Elmhurst, Ill. 60126

*Business Management*
22 W. Putnam Ave.
Greenwich, Conn. 06830

*Car Craft*
5959 Hollywood Blvd.
Los Angeles, Calif. 90028

*Car Life*
1499 Monrovia Ave.
Newport Beach, Calif. 92663

*Ceramic Age*
2800 Euclid Ave.
Cleveland, Ohio 44115

*Chemical Engineering*
330 W. 42nd St.
New York, N.Y. 10036

*Chemical & Engineering News*
1155 16th St. N.W.
Washington, D.C. 20036

*Chemical Engineering Progress*
345 E. 47th St.
New York, N.Y. 10017

*Chemical Processing*
111 E. Delaware Pl.
Chicago, Ill. 60611

*Civil Engineering*
345 E. 47th St.
New York, N.Y. 10017

*Coal Age*
330 W. 42nd St.
New York, N.Y. 10036

*Coal Mining & Processing*
300 W. Adams St.
Chicago, Ill. 60606

*College Management*
22 W. Putnam Ave.
Greenwich, Conn. 06830

*Color Engineering*
18 John St.
New York, N.Y. 10038

*Combustion*
277 Park Ave.
New York, N.Y. 10017

*Commercial Car Journal*
Chestnut & 56th Sts.
Philadelphia, Pa. 19139

*Communications News*
402 W. Liberty Dr.
Wheaton, Ill. 60817

*Compressed Air Magazine*
942 Memorial Pkwy.
Phillipsburg, N.J. 08865

*Computer Design*
Professional Bldg.
Baker Ave.
West Concord, Mass. 01781

*Computers and Automation*
815 Washington St.
Newtonville, Mass. 02160

*Computerworld*
60 Austin St.
Newton, Mass. 02160

*Concrete Construction*
Box 555
Elmhurst, Ill. 60126

*Concrete Products*
300 W. Adams St.
Chicago, Ill. 60606

*Construction Equipment & Materials*
205 E. 42nd St.
New York, N.Y. 10017

*Construction Machinery Maintenance*
Box 202
Barrington, Ill. 60010

*Construction Methods & Equipment*
330 W. 42nd St.
New York, N.Y. 10036

*Constructor*
1957 E. St. N.W.
Washington, D.C. 20006

*Consulting Engineer*
217 Wayne St.
St. Joseph, Mich. 49805

*Contemporary Design*
1202 S. Park St.
Madison, Wis. 53715

*Contractors & Engineers*
Berkshire Commons
Pittsfield, Mass. 01201

*Control Engineering*
466 Lexington Ave.
New York, N.Y. 10017

*Co-Operative Builder (Farm)*
739 Johnson St. N.E.
Minneapolis, Minn. 55413

*Cotton*
416 Hickman Bldg.
Memphis, Tenn. 38103

*Cryogenic Engineering News*
2800 Euclid Ave.
Cleveland, Ohio 44115

*Data Systems News*
200 Madison Ave.
New York, N.Y. 10016

*Design News*
270 St. Paul St.
Denver, Colo. 80206

*Die and Stamping News*
15936 Kinloch Rd.
Detroit, Mich. 48237

*Diesel & Gas Turbine Progress*
Box 7046
Milwaukee, Wis. 53213

*Distribution Manager*
56th & Chestnut Sts.
Philadelphia, Pa. 19139

*Domestic Engineering*
1801 Prairie Ave.
Chicago, Ill. 60616

*Dun's Review*
Box 3088, Grand Central Station
New York, N.Y. 10017

*EEE*
820 Second Ave.
New York, N.Y. 10017

*Electric Heat & Air Conditioning*
447 Orange Rd.
Montclair, N.J. 07042

*Electric Heating Journal*
3132 Fordem Ave.
Madison, Wis. 53701

*Electric Light & Power*
221 Columbus Ave.
Boston, Mass. 02116

*Electrical Wholesaling*
330 W. 42nd St.
New York, N.Y. 10036

*Electrical World*
330 W. 42nd St.
New York, N.Y. 10036

*Electromechanical Design*
167 Corey Road
Brookline, Mass. 02146

*Electronic Capabilities*
209 Dunn Ave.
Stamford, Conn. 06905

*Electronic Design*
850 Third Ave.
New York, N.Y. 10022

*The Electronic Engineer*
56th & Chestnut Sts.
Philadelphia, Pa. 19139

*Electronic Instrument Digest*
222 W. Adams St.
Chicago, Ill. 60606

*Electronic Packaging and Production*
222 W. Adams St.
Chicago, Ill. 60606

*Electronic Products*
645 Stewart Ave.
Garden City, N.Y. 11530

*Electronics*
330 W. 42nd St.
New York, N.Y. 10036

*Electronics World*
One Park Ave.
New York, N.Y. 10016

*Electro-Technology*
Industrial Research Bldg.
Beverly Shores, Ind. 46301

*Engineer*
345 E. 47th St.
New York, N.Y. 10017

*Engineering Graphics*
25 W. 45th St.
New York, N.Y. 10036

*Engineering & Mining Journal*
330 W. 42nd St.
New York, N.Y. 10036

*Engineering News-Record*
330 W. 42nd St.
New York, N.Y. 10036

*Environmental Science & Technology*
1155 16th St. N.W.
Washington, D.C. 20036

*Evaluation Engineering*
1282 Old Skokie Rd.
Highland Park, Ill. 60035

*Farm Technology*
37841 Euclid Ave.
Willoughby, Ohio 44094

*Fleet Owner*
330 W. 42nd St.
New York, N.Y. 10036

*Food & Drug Packaging*
777 Third Ave.
New York, N.Y. 10017

*Food Engineering*
Chestnut & 56th Sts.
Philadelphia, Pa. 19139

*Food Ingredients & Equipment*
111 E. Delaware Pl.
Chicago, Ill. 60611

*Food Plant Ideas*
731 Hennepin Ave.
Minneapolis, Minn. 55403

*Food Processing*
111 E. Delaware Pl.
Chicago, Ill. 60611

*Food Technology*
221 N. LaSalle St.
Chicago, Ill. 60601

*Foundry*
1213 W. Third St.
Cleveland, Ohio 44113

*Geotimes*
1444 N. St. N.W.
Washington, D.C. 20005

*Graphic Science*
9 Maiden La.
New York, N.Y. 10038

*GSE (Ground Support Equipment)*
3110 Mt. Vernon Ave.
Alexandria, Va. 22305

*Handling & Shipping*
614 Superior Ave. W.
Cleveland, Ohio 44113

*Hard Goods & Soft Goods Packaging*
777 Third Ave.
New York, N.Y. 10017

*Harvard Business Review*
Soldiers Field
Boston, Mass. 02163

*Heating, Piping & Air Conditioning*
10 S. LaSalle St.
Chicago, Ill. 60603

*Journal of Home Economics*
1600 20th St. N.W.
Washington, D.C. 20009

*House & Home*
330 W. 42nd St.
New York, N.Y. 10036

*Hydraulics & Pneumatics*
614 Superior Ave. W.
Cleveland, Ohio 44113

*Hydrocarbon Processing*
Box 2608
Houston, Tex. 77001

*IEEE Spectrum*
345 E. 47th St.
New York, N.Y. 10017

*Illuminating Engineering*
345 E. 47th St.
New York, N.Y. 10017

*Industrial Design*
18 E. 50th St.
New York, N.Y. 10022

*Industrial Distribution*
330 W. 42nd St.
New York, N.Y. 10036

*Journal of Industrial Engineering*
345 E. 47th St.
New York, N.Y. 10017

*Industrial & Engineering Chemistry*
1155 16th St. N.W.
Washington, D.C. 20036

*Industrial Finishing*
Hitchcock Bldg.
Wheaton, Ill. 60187

*Industrial Heating*
1400 Union Trust Bldg.
Pittsburgh, Pa. 15219

*Industrial Improvement*
20 N. Wacker Dr.
Chicago, Ill. 60606

*Industrial Maintenance & Plant Operation*
One W. Olney Ave.
Philadelphia, Pa. 19120

*Industrial Marketing*
740 Rush St.
Chicago, Ill. 60611

*Industrial Photography*
200 Madison Ave.
New York, N.Y. 10016

*Industrial Research*
Beverly Shores, Ind. 46301

*Industrial Water Engineering*
437 Madison Ave.
New York, N.Y. 10022

*Instrument and Control Systems*
56th & Chestnut Sts.
Philadelphia, Pa. 19139

*Instrumentation Technology*
530 William Penn Pl.
Pittsburgh, Pa. 15219

*International Management*
330 W. 42nd St.
New York, N.Y. 10036

*The Iron Age*
56th & Chestnut Sts.
Philadelphia, Pa. 19139

*Laboratory Management*
200 Madison Ave.
New York, N.Y. 10016

*Machine Design*
1213 W. Third St.
Cleveland, Ohio 44113

*Machinery*
200 Madison Ave.
New York, N.Y. 10016

*Management Review*
135 W. 50th St.
New York, N.Y. 10020

*Material Handling Engineering*
614 Superior Ave. W.
Cleveland, Ohio 44113

*Materials Engineering*
430 Park Ave.
New York, N.Y. 10022

*Materials Protection*
2400 W. Loop S.
Houston, Tex. 77027

*Mayor & Manager*
5811 Dempster
Morton Grove, Ill. 60053

*Mechanical Engineering*
345 E. 47th St.
New York, N.Y. 10017

*Metal Progress*
Metals Park, Ohio 44073

*The Magazine of Metals Producing*
24 Commerce St.
Newark, N.J. 07102

*Metalworking*
221 Columbus Ave.
Boston, Mass. 02116

*Metfax Magazine*
13601 Euclid Ave.
Cleveland, Ohio 44112

*Microwave Journal*
610 Washington St.
Dedham, Mass. 02026

*Microwaves*
850 Third Ave.
New York, N.Y. 10022

*The Military Engineer*
800 17th St. N.W.
Washington, D.C. 20006

*Minerals Processing*
380 Northwest Hwy.
Des Plaines, Ill. 60016

*Mining Engineering*
345 E. 47th St.
New York, N.Y. 10017

*Modern Casting*
Golf & Wolf Rds.
Des Plaines, Ill. 60016

*Modern Chemicals*
27 N. Ward Ave.
Rumson, N.J. 07760

*Modern Concrete*
105 W. Adams St.
Chicago, Ill. 60603

*Modern Converter*
Ojibway Bldg.
Duluth, Minn. 55802

*Modern Data Systems*
3 Lockland Ave.
Framingham, Mass. 01701

*Modern Government*
10 River St.
Stamford, Conn. 06904

*Modern Machine Shop*
600 Main St.
Cincinnati, Ohio 45202

*Modern Manufacturing*
330 W. 42nd St.
New York, N.Y. 10036

*Modern Materials Handling*
221 Columbus Ave.
Boston, Mass. 02116

*Modern Metals*
433 N. Michigan Ave.
Chicago, Ill. 60611

*Modern Packaging*
1301 Avenue of the Americas
New York, N.Y. 10019

*Modern Plastics*
1301 Avenue of the Americas
New York, N.Y. 10019

*Modern Railroads*
5 S. Wabash Ave.
Chicago, Ill. 60603

*Modern Textiles Magazine*
303 Fifth Ave.
New York, N.Y. 10016

*Motor Age*
Chestnut & 56th Sts.
Philadelphia, Pa. 19139

*National Engineer*
National Association of Power Engineers
176 W. Adams St.
Chicago, Ill. 60603

*National Petroleum News*
330 W. 42nd St.
New York, N.Y. 10036

*National Public Accountant*
1717 Pennsylvania Ave.
Washington, D.C. 20006

*National Safety News*
425 N. Michigan Ave.
Chicago, Ill. 60611

*Nation's Cities*
1612 K St. N.W.
Washington, D.C. 20006

*Nation's Schools*
1050 Merchandise Mart
Chicago, Ill. 60654

*Nuclear News*
244 E. Ogden Ave.
Hinsdale, Ill. 60521

*Oil and Gas Equipment*
Box 1260
Tulsa, Okla. 74101

*The Oil and Gas Journal*
Box 1260
Tulsa, Okla. 74101

*Ordnance*
616 Transportation Bldg.
Washington, D.C. 20006

*Package Engineering*
2 N. Riverside Plaza
Chicago, Ill. 60606

*Paper Age*
466 Kinderkamack Rd.
Oradell, N.J. 07649

*Paper, Film & Foil Converter*
200 S. Prospect Ave.
Park Ridge, Ill. 60068

*Paper Trade Journal*
551 Fifth Ave.
New York, N.Y. 10017

*Park Maintenance*
Box 409
Appleton, Wis. 54911

*Parks and Recreation*
1700 Pennsylvania Ave. N.W.
Washington, D.C. 20006

*Personnel Journal*
100 Park Ave.
Swarthmore, Pa. 19081

*Petro/Chem Engineer*
Box 1589
Dallas, Tex. 75221

*The Petroleum Engineer*
Box 1589
Dallas, Tex. 75221

*Photo Methods for Industry*
33 W. 60th St.
New York, N.Y. 10023

*Photographic Applications in Science and Technology*
41 E. 28th St.
New York, N.Y. 10016

*Pipe Line Industry*
Box 2608
Houston, Tex. 77001

*Pipe Line News*
1217 Kennedy Blvd.
Bayonne, N.J. 07002

*The Pipeline Engineer*
Box 1589
Dallas, Tex. 75221

*Pit and Quarry*
105 W. Adams St.
Chicago, Ill. 60603

*Plant Engineering*
308 E. James St.
Barrington, Ill. 60010

*Plant Operating and Management*
205 E. 42nd St.
New York, N.Y. 10017

*Plant Services*
111 E. Delaware Pl.
Chicago, Ill. 60611

*Plastics Design and Processing*
311 E. Park Ave.
Libertyville, Ill. 60048

*Plastics Technology*
630 Third Ave.
New York, N.Y. 10017

*Plastics World*
270 St. Paul St.
Denver, Colo. 80206

*The Journal of Plumbing, Heating and Air Conditioning*
92 Martling Ave.
Tarrytown, N.Y. 10591

*Plumbing Heating Cooling Business*
1016 20th St. N.W.
Washington, D.C. 20036

*Power*
330 W. 42nd St.
New York, N.Y. 10036

*Power Engineering*
1301 S. Grove Ave.
Barrington, Ill. 60010

*Power Transmission Design*
614 Superior Ave. W.
Cleveland, Ohio 44113

*Precision Metal*
614 Superior Ave. W.
Cleveland, Ohio 44113

*Production Equipment*
407 S. Dearborn St.
Chicago, Ill. 60605

*Production Magazine*
Box 101
Bloomfield Hills, Mich. 48013

*Products Finishing*
600 Main St.
Cincinnati, Ohio 45202

*Professional Engineer*
2029 K St. N.W.
Washington, D.C. 20006

*Public Management*
1140 Connecticut Ave. N.W.
Washington, D.C. 20036

*Public Power*
2600 Virginia Ave. N.W.
Washington, D.C. 20037

*Public Utilities Fortnightly*
425 13th St. N.W.
Washington, D.C. 20004

*Public Works Magazine*
200 S. Broad St.
Ridgewood, N.J. 07451

*Pulp & Paper*
370 Lexington Ave.
New York, N.Y. 10017

*Quality Assurance*
Hitchcock Bldg.
Wheaton, Ill. 60187

*Quality Progress*
161 W. Wisconsin Ave.
Milwaukee, Wis. 53203

*Quick Frozen Foods*
205 E. 42nd St.
New York, N.Y. 10017

*Railroad Magazine*
205 E. 42nd St.
New York, N.Y. 10017

*Railway Age*
30 Church St.
New York, N.Y. 10007

*Railway Signaling and Communications*
14 E. Jackson Blvd.
Chicago, Ill. 60604

*Railway Track and Structures*
14 E. Jackson Blvd.
Chicago, Ill. 60604

*R/D Research Development*
205 W. Wacker Dr.
Chicago, Ill. 60606

*Reprographics*
200 Madison Ave.
New York, N.Y. 10016

*Roads and Streets*
209 W. Jackson Blvd.
Chicago, Ill. 60606

*Rock Products*
300 W. Adams St.
Chicago, Ill. 60606

*Rural Electrification*
2000 Florida Ave. N.W.
Washington, D.C. 20009

*Rural and Urban Roads*
209 W. Jackson Blvd.
Chicago, Ill. 60606

*SAE Journal*
Two Pennsylvania Plaza
New York, N.Y. 10001

*Sales/Marketing Today*
630 Third Ave.
New York, N.Y. 10017

*SAMPE Journal*
647 N. Sepulveda, Bel Air
Los Angeles, Calif. 90049

*School Management*
22 W. Putnam Ave.
Greenwich, Conn. 06830

*Science*
1515 Massachusetts Ave. N.W.
Washington, D.C. 20005

*Science Digest*
575 Lexington Ave.
New York, N.Y. 10022

*Scientific American*
415 Madison Ave.
New York, N.Y. 10017

*Scientific Research*
330 W. 42nd St.
New York, N.Y. 10036

*Software Age*
1020 Church St.
Evanston, Ill. 60201

*Solid State Technology*
14 Vanderventer Ave.
Port Washington, N.Y. 11050

*Solid Wastes Management*
150 E. 52nd St.
New York, N.Y. 10022

*Space/Aeronautics*
205 E. 42nd St.
New York, N.Y. 10017

*Space Technology International*
425 National Press Bldg.
Washington, D.C. 20004

*Space World*
Amherst, Wis. 54406

*Steelways*
150 E. 42nd St.
New York, N.Y. 10017

*Supervisory Management*
135 W. 50th St.
New York, N.Y. 10020

*Telecommunications*
610 Washington St.
Dedham, Mass. 02026

*Telephone Engineer & Management*
402 W. Liberty Dr.
Wheaton, Ill. 60187

*Telephony*
53 W. Jackson Blvd.
Chicago, Ill. 60604

*Test Engineering & Management*
61 Monmouth Rd.
Oakhurst, N.J. 07755

*Textile Equipment*
Box 12415
Charlotte, N.C. 28205

*Textile Industries*
1760 Peachtree Rd. N.W.
Atlanta, Ga. 30309

*Textile World*
330 W. 42nd St.
New York, N.Y. 10036

*Today's Transport International*
Box 1256
Stamford, Conn. 06904

*The Tool and Manufacturing Engineer*
20501 Ford Rd.
Dearborn, Mich. 48128

*Tooling & Production*
13601 Euclid Ave.
Cleveland, Ohio 44112

*Traffic Engineering*
1725 DeSales St. N.W.
Washington, D.C. 20036

*Traffic Safety*
425 N. Michigan Ave.
Chicago, Ill. 60611

*Trains*
1027 N. 7th St.
Milwaukee, Wis. 53233

*Transmission & Distribution*
One River Rd.
Cos Cob, Conn. 06807

*Transport Topics*
1616 P St. N.W.
Washington, D.C. 20036

*Transportation & Distribution Management*
815 Washington Bldg.
Washington, D.C. 20005

*Undersea Technology*
1117 N. 19th St.
Arlington, Va. 22209

*Visual Communications Instructor*
25 W. 45th St.
New York, N.Y. 10036

*Water & Sewage Works*
35 E. Wacker Dr.
Chicago, Ill. 60601

*Water & Wastes Engineering*
466 Lexington Ave.
New York, N.Y. 10017

*What's New in Chemical Processing Equipment*
111 E. Delaware Pl.
Chicago, Ill. 60611

*What's New in Home Economics*
466 Lexington Ave.
New York, N.Y. 10017

*World Construction*
466 Lexington Ave.
New York, N.Y. 10017

*World Oil*
Box 2608
Houston, Tex. 77001

*World Mining*
500 Howard St.
San Francisco, Calif. 94105

*World Petroleum*
25 W. 45th St.
New York, N.Y. 10036

# INDEX